Hinges

Hinges

Meditations on the Portals of the Imagination

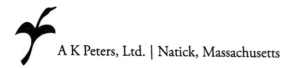

Grace Dane Mazur

A K Peters, Ltd. | Natick, Massachusetts

Editorial, Sales, and Customer Service Office

A K Peters, Ltd.
5 Commonwealth Road, Suite 2C
Natick, MA 01760
www.akpeters.com

A portion of an earlier version of Chapter I appeared in *The Envisioned Life: Essays in Honor of Eva Brann*, edited by Peter Kalkavage and Eric Salem (Paul Dry Books, Philadelphia, 2007).

Library of Congress Cataloging-in-Publication Data

Mazur, Grace Dane.
  Hinges : meditations on the portals of the imagination/ Grace Dane Mazur.
      p. cm.
  Includes bibliographical references.
  ISBN 978-1-56881-715-6 (alk. paper)
  1. Authorship--Philosophy. 2. Death. 3. Death in literature. I. Title.
  PS3563.A9872H56 2010
  818'.54--dc22
                                  2010012515

Cover by Josef Beery. Background: "The American Cowslip" by Dr. Robert John Thornton; photograph by Paul Horowitz. Inset: *Girl Asleep* (*Fille Assoupie*) by Johannes Vermeer; photograph by Ana Dane. Author photograph by Jim Harrison.

Printed in India
14 13 12 11 10   10 9 8 7 6 5 4 3 2 1

For Barry, Zeke, and Sasha

Table of Contents

*Tum demum horrisono stridentes cardine sacrae
panduntur portae.*

                              The awaited
Time has come, hell gates will shudder wide
On shrieking hinges.

                           —Virgil, *The Aeneid*

Entrance

One of the great mysteries is the altered state that writing brings us to: the way we cross over—when we read or write—from the "real" world that normally surrounds and claims us, into the imagined world of fiction or poetry. Strange and marvelous things happen when we are captured and captivated in this way. The mind's eye becomes more powerful than the body's eye; time changes color; language undergoes a shift so that we seem to speak a different native tongue.

For a long time, I have been obsessed with boundaries and margins, thresholds, and doors—both figurative and literal. I love the way doors open to a new room, a new thought, and how it is their hinges that allow us to swing them with ease or trepidation. In certain cultures one does not step on the door sill, as spirits reside there.

This book is not a scholarly discourse but rather a series of meditations on different facets of the entrances and entrancements of writing. In examining the threshold between real and imagined worlds, I consider how our own passage to the world of the imagination echoes the voyage of the hero crossing to the Other World, or descending to the Land of the Dead. How necessary such crossings are and how odd the portals, with their glorious hardware.

To think about these transitions can also be, in my view, to enact the transformation itself. Thus at times I explore at greater length a text (such as Katherine Mansfield's "The Garden Party" or Eudora Welty's "Music from Spain"); a visual image (such as *Orpheus and Eurydice* by Peter Paul Rubens); a concept (such as Forbidden Looking); or even, in the case of the hinge, actual hardware. These analyses take on a life of their own.

In Chapter I, I start by discussing the hardware of the threshold—physical and metaphysical—hinges, axles, and trapdoors. Why do important hinges have to make hideous and unearthly noises?

In Chapter II, after considering the shakiness at the threshold of fictional narratives, we proceed to the Other World and the Land of the Dead to look at those who explore there, why they go, and what they find. I tell of my own descent in the caves at Lascaux as a preface to stories of archaic heroes—Gilgamesh, Parmenides, and Odysseus—as well as the modern heroine of Katherine Mansfield's "The Garden Party," all of whom cross to that Other World and triumphantly return.

In Chapter III, we look at the case of Orpheus, a less triumphant visitor to the Land of the Dead, and we consider the nature of looking and seeing, why looking is sometimes forbidden, and why the gaze is so important that it can be fatal.

In Chapter IV, we return to Hell, and one of its most successful heroes, Christ, who goes there to release the prophets and patriarchs of the Old Testament. We go more deeply into questions of the hinges of Hell, as well as hinges of the body, mind, and heart. The hinges of the day present themselves, along with the peculiar moods that befall apes and humans at dusk, and gods at the twilight of creation. Then we look at the nested trapdoors and oscillations in Eudora Welty's brilliant story "Music from Spain."

In Chapter V, we look at hinges of the mind and of the heart in hopes of seeing how these rotating and pivoting instruments cast light on form and content in fiction and poetry, and on the creative process itself.

The texts and images included here range over a span of seventeen thousand years, so I have included a rough time-line at the end of the book for chronological convenience.

Acknowledgments

What a joy, finally, to thank friends for their inspiration, conversation, and guidance. Barry Mazur, Zeke Mazur, and Sasha Makarova act as continual sources of thought-ignition and help me with the fires that ensue. Eva Brann, Jamaica Kincaid, Milen Poenaru, Valentin Poenaru, Sue Trupin, Jeffrey Levine, Greg Nagy, Elaine Scarry, Natasha Bershadsky, Cecie and Paul Dry, Chris Nelson, Joyce Olin, Milenko Matanovic, and Kathi Lightstone all have delighted me with their insights, and many have listened to me at great length and given wise counsel.

Philip Fisher first spoke to me of thresholds and enchantment in a chance meeting in Harvard Yard many years ago. From that conversation came the seed of this work, and I am infinitely grateful.

This book grew out of correspondence over the years with Jeffrey Levine, whose writings remain a continual source of inspiration. Early drafts of each chapter became lectures for the residencies and alumni conferences of the MFA Program for Writers at Warren Wilson. I am grateful to my faculty colleagues, fellow alumni, and students for the most intense, demanding, and enlightening dialogue I could ever hope for.

For crucial pieces of information, references, suggestions, and other sorts of help I thank Avner Ash, Rosemary Bernard, Laura Hendrie, Suzanne Koven, Tony Millham, Stelios Moschapidakis, Mark Solomon, Nell Stenson, and Robin Tzannes.

For help with the images—locating, procuring, photographing, and translating—my thanks to Natasha Bershadsky, Capi Corrales, Ana Dane, Paul Horowitz, Sasha Makarova, and Dorina Papaliou.

I am grateful for Paul Dry's wise and generous commentaries throughout the course of this project. Barbara Ras provided superb insights and wonderful advice on several versions of this manuscript. I am much indebted to her, and to Erwin Cook, whose scholarly comments on the classical material in an early draft were greatly needed and much appreciated. Any mistakes that remain are stubbornly my own.

I am immensely grateful to Klaus and Alice Peters for the depth and breadth of their vision as publishers and editors. Their intellect, energy, and generosity are astonishing and wonderful. It has been a great pleasure to work with Charlotte Henderson, whose edits are meticulous and whose superb questions always penetrate to the heart of the matter. And finally, my thanks to Camber Agrelius, who turned these pages into a book.

I | The Hinges of Hell

Introduction to the Hinge

Picture the innocent honeymoon couple of the horror movie as they sit in their car that has just died, leaving them marooned in the storm on the deserted country road. Finally, they run through the flailing rain to the mansion on the hill; they ring the doorbell and the doors open, slowly, with an ominous rising shriek. We know these doors will shriek. There is no other way.

But why, in fact, are the doors so loud?

Actually, I am wrong. The doors do not sound; it is the hinges that sing out in the night. Metal against metal, metal against stone. Milton knows this: in *Paradise Lost*, he tells how Sin, the daughter and mistress of Satan, agrees to open the gates of hell:

> . . . then in the key-hole turns
> Th'intricate wards, and every bolt and bar
> Of massy iron or solid rock with ease
> Unfastens; on a *sudden*, open fly
> With impetuous recoil and jarring sound
> Th'infernal doors, and on their hinges grate
> Harsh thunder, that the lowest bottom shook
> Of Erebus . . .

> [BOOK II, LINES 871–883]

Here, Milton's syntax feels as marvelously grating as the sound he is describing. But the fact that something strange was going on with these particular hinges—the hinges of the gates of Hell—was also known to the artists of the Renaissance and the Byzantine world.

There is a quiet fresco by Fra Angelico in the museum of San Marco in Florence. Christ, in his radiant halo, floats just above the ground inside the doorway to the cavern of Hell, reaching toward a group of haloed figures in immaculate robes. This is *Christ Harrowing Hell*, also called *Christ in Limbo* (Figure 1).

Figure 1. *Christ in Limbo* by Fra Angelico (see Plate I).

The composition of the Fra Angelico fresco is well balanced, with daylight and Christ on the right, demons fleeing or crouching high in the shadows on the left, and the crowd of prophets in the middle. But what caught my eye, and led to one of the obsessions at the core of this book, is the black hinge on the doorframe (more precisely, a *pintle* hinge). Pictorially there is no need for this weird little object in the composition; nor is there any need for the detailed depiction of the corresponding hinge straps on the door. (See Figure 2.) The lock has been ripped off; it lies twisted at the top of the door; a couple of nails the size of crucifixion nails are strewn about. Though Christ hovers just above the ground with his flag, his entrance has been violent enough to smash down the door and open sharp new cracks in the floor of Hell.

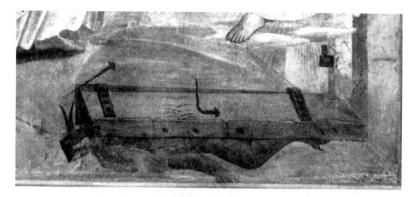

Figure 2. Detail of *Christ in Limbo* by Fra Angelico.

A pattern begins to emerge when we look at how Albrecht Dürer treats the same situation. In each of Dürer's *Passions*, Christ stands in Limbo holding a tall staff with a cross on its banner, a sign of his victory over Hell. He has just pulled out Adam and Eve, who are naked, and he is reaching into the arched entrance of Hell for somebody else. In *The Engraved Passion*, Moses—recognizable by his horns and by a corner of his tablets just visible behind Adam's arm—stands at the extreme left, having already been pulled from Hell.* Demons abound. The broken door with its prominent hinge has fallen outside Hell. In both *The Large Passion* (Figure 3) and *The Engraved Passion* (Figure 4), the hinge is prominent in the lower left. While both images are astonishing, in *The Engraved Passion* something supremely strange is going on with the point of view: we see the backs of the heads of those who are being liberated by Christ. The only possible interpretation of this view is that we ourselves are located even deeper in the cavern of Hell.

Back to my original question: What is going on with these hinges? Why do they howl and shriek in certain situations? Why are certain artists so painstaking in their depiction of them?

---

*The idea that Moses, and by extension all Jews, had horns comes from Exodus 34:30–31. Moses, who has just spent forty days and forty nights in god's presence, is described as having a fiery radiance about him: "and, look, the skin of his face glowed." Here the word for "glow" is *qaran*, which occurs only at this point in the Hebrew Bible. The Greek and Latin translations substituted the word *qeren*, meaning "horns." See Robert Alter, *The Five Books of Moses*, page 512, footnote 9.

Hardware and Definitions

In what follows I will argue that hinges, miraculous objects, are everywhere in writing: in poetry and fiction—in form as well as content—and in the process itself. And that they are crucial not only in writing but in all other important aspects of life as well.

But first let us consider the actual hardware, the old-fashioned metal hinges that we find in a hardware shop, or have made by a blacksmith. According to the *Oxford English Dictionary*, a *hinge* is the movable

Figure 3. *Harrowing of Hell* from *The Large Passion* by Albrecht Dürer. (Photograph © 2010 Museum of Fine Arts, Boston.)

Figure 4. *Harrowing of Hell* from *The Engraved Passion* by Albrecht Dürer. (Photograph © 2010 Museum of Fine Arts, Boston.)

joint or mechanism by which a gate or door is hung upon the side-post, so as to be opened or shut by being turned upon this joint.*

An *axle* is the center-pin or spindle upon which a wheel revolves, or which revolves along with it. It is also, in obsolete usage, the imaginary line about which a planet or the sky or the heavens revolve.

A hinge is really a restricted case of an axle. Instead of making full revolutions of 360 degrees, the hinge plate with its attached door simply pivots back and forth, generally through about 90 degrees.

---

* "Hotter than the Hinges of Hell" is a common phrase from my childhood. The physics of this situation are clear: Hinges are generally made of metal, and so, whether the infernal doors are wood, as they often are in art, or metal, as Milton describes them, the metal hinges are going to conduct the heat of the forbidden hellfires within, metal being a superb conductor of heat.

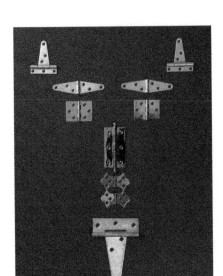

Figure 5. Some common hinges.

A *revolving door* is more like a wheel turned on its side. It has a central vertical axle instead of a hinge pin, and we insert ourselves between the spokes, which are extended into tall vertical planes.

The door that is not plumb, not correctly suspended from its hinges, is like a carcass, a side of beef, dead weight; it is pretty useless. It can fall open, but not swing shut. This is why becoming *unhinged* is such a serious thing. You collapse wildly; you swing heavily askew.

One form that becoming unhinged can take is *obsession*. Although the etymology of *obsession* implies that something sits on us or besieges us—from the Latin *ob* meaning against, toward, over, and *sedere*, to sit—perhaps one can also think of it as when we sit in one of the rooms of our mind, unable to perform the hinging action to take us to any other room.

Both hinges and axles generally allow easy motion back and forth. Whenever we attempt to drag a chariot with no wheels, or to lift a door, particularly an old door of solid oak or cherry, it becomes obvious how miraculous the pivoting rotary motion of axles and hinges is. In the case of the door, the surprise is that so few metal plates, together with a small handful of screws, can not only hold up the much heavier door, but also allow it to swing so effortlessly that it appears to float.

This lack of effort with the hinged vertical door is what makes the horizontal *trapdoor* so different. Although some trapdoors are pried open and lifted in order to enter the space below, the ones I am interested in are snares, hidden or invisible traps that open precipitously when you step on them, delivering you to the depths. Then tremendous energy is needed to climb back up to where you started.

A bolt or lock generally works at right angles to the hinge and functions to stop in its tracks the very movement that the hinge so beautifully allows. There are some bolts that only divinities can work: Satan's incestuous mistress, Sin, is the only one who can open the Gates of Hell as in the passage from *Paradise Lost* quoted earlier. Hermes, who often shuttles between the living and the dead, is the only one who can slide the heavy bolt leading to a place of death in Book 24 of *The Iliad*.

Hinges seem to go all the way back in human history; the ancient Sumerians had them, once they moved from round daub and wattle huts to rectangular houses made of clay brick. Sumerian doors had a pair of vertical pins at the hinging corners. The bottom pin rested in a stone door socket, usually the only architectural stone in the house. The upper pin was held in place by a strap, and the door frame was painted red to scare away demons.[1]

❧ ❦

Given the beautiful working of hinges, the almost magical ease and apparent weightlessness that they bring about, what accounts for the horrifying noises when Sin opens the gates of Hell, or when the young couple in the horror movie open the door of the terrible mansion? I think they all have the same cause: crucial changes of state demand and deserve audible markers. The cry of impending arrival. Birth pangs. The howl of the newborn. Sex moans. Ecstasies. Death rattles. What we have here is clearly the groan of the liminal.

This alarm at the threshold signals the importance of the transition taking place, and alerts us to the transformations that will result. But there is a related form of the liminal that happens when we are alone with a text. This one is quieter, without trumpet or fanfare, and equally important. It has to do with reading and writing our way into the imagined world.

II | The World of Fiction and the Land of the Dead

Thresholds and Instabilities

We curl up with a novel. The very verb implies a narrowing, a spiraling inward of our attention. If all goes well, we soon enter an altered state of consciousness, in which we descend into a world not our own. What is the nature of these other worlds? What do we find when we get there?

Stories begin with instabilities—perhaps because beginnings themselves are such unstable conditions. In fact, the opening pages sometimes show the protagonist in a condition of both liminality and entrancement, liminality being the state of being on the threshold. It is as though there is a sense of, "Look, reader, the same thing that is happening to you—now that you are coiled around this book and are about to slip into the imagined world—is happening to this fictional character, who is at the edge of his own altered consciousness, and at the edge of adventure."[1]

Two pre-modern works that open with an extreme sense of liminality and entrancement are Proust's *Remembrance of Things Past* and Lewis Carroll's *Alice in Wonderland*. Proust's Overture in *Swann's Way* begins: "*Longtemps je me couchais de bonne heure* (For a long time I would go to bed early)." The next fifty pages are a breathtaking meditation on going to bed, falling asleep, waking too early, drowsing. Not only is Marcel at the border of sleeping and waking, but his memories hover at the very edge of coming into view.

Similarly, Alice at the beginning of her adventure drowses "sleepy and stupid" in the heat, wondering

> whether the pleasure of making a daisy-chain would be worth the
> trouble of getting up and picking the daisies, when suddenly a White
> Rabbit with pink eyes ran close by her. . . . She ran across the field after

it, and fortunately was just in time to see it pop down a large rabbit-hole under the hedge. In another moment down went Alice after it, never once considering how in the world she was to get out again.[2]

The openings of some contemporary American masterpieces show the same sort of liminality and entrancement and also an intricate imbalance leading to a sort of structural instability—that state where things are so precarious that something has got to happen. This structural instability can come from being on the edge, or simply being *on edge*, and is often accompanied by uneasiness, excitement, fear. Often *time* is a preoccupation as well. As an example of what I mean, consider the first few sentences of Paula Fox's novel *The Widow's Children*:

> Clara Hansen, poised upright in her underwear on the edge of a chair, was motionless. Soon she must turn on a light. Soon she must finish dressing. She would permit herself three more minutes in her darkening apartment in that state that was so nearly sleep. She turned to face a table on which sat a small alarm clock.

Clara is "poised," the verb implies the necessity of action or change—she is on the edge of her chair, upright, not quite on the chair, not off. Not asleep, but nearly; neither dressed nor naked, in the apartment that is neither light nor dark—but getting darker—on an afternoon that is turning into evening.

> At once, a painful agitation brought her to her feet. She would be late: buses were not reliable. She could not afford a taxi to take her to the hotel where her mother, Laura, and Laura's husband, Desmond Clapper, were expecting her for drinks and dinner. In the morning the Clappers were sailing away on a ship—this time, to Africa. They would be gone for months. Clara had managed to get away from the office where she worked a half hour early so she would have time enough. But it had been time enough to fall into a dream of nothingness.

Clara is at the edge of sleep, in a dream of nothingness, but also in a turmoil over the notion of Time, and time words are ticking like some furious vexed clock all through these opening pages: "Soon," "soon," "three more minutes," "alarm clock," "at once," "late," "in the morning," "this time," "months," "time enough," "time enough," "evening," "evening," "early April," "still," "evening," "long ago."

> Clara went quickly to the small bedroom where her dress lay across the bed. It was the best thing she owned. She was aware that as a rule she dressed defensively. But she had made a perverse choice for this evening. Laura would know the dress was expensive. The hell with it, she told herself, but felt only irresoluteness as the silk settled against her skin.

Clara's silk dress is inappropriate, too expensive for the occasion, and she gets none of the sensual pleasure that one does from wearing silk, but feels only her overriding psychic state of irresoluteness. Later, however, she counteracts the glorious luxury of her dress with a raincoat that is too shabby and dirty for the occasion, and hurries herself into a self-induced state of "distressed excitement."

> A few drops of rain slid down the windows as she passed through the living room. She turned on a light to come home to, and for a brief moment, it seemed the evening was already over, that she had returned, consoled by the knowledge that once Laura was gone, she hardly need think of her. After all, the occasions of their meetings were so rare.

The first rain drops have just appeared, signs of the leading edge of the storm. Both the weather and the season are on edge—that cruelty of late spring when the air should be warm but neglects to be. Finally Clara does something so strange and lonely: she turns on "a light to come home to." This preemptive strike against the dark shows her solitary existence: *she* is the one who greets her when she comes home, she provides light and warmth—but notice also how she is looping together two separate points in time, before and after seeing her difficult mother, Laura, as though, if she could, she would loop that interaction out of existence.

So many instabilities—postural, psychological, diurnal, seasonal, meteorological—and we read on, sucked in by all the teetering, waiting for something to topple or crash.

                                                   ෨ ෫

While the opening of Fox's *The Widow's Children* shows Clara dozing when she should be waking, finally transforming her rigid calm into excited distress, in the beginning of Charles Baxter's

The Feast of Love, we find our narrator, named Charlie Baxter, waking in a panic in the middle of the night.

> The man—me, this pale being, no one else, it seems—wakes in fright, tangled up in the sheets.
>
> The darkened room, the half-closed doors of the closet and the slender pine-slatted lamp on the bedside table: I don't recognize them. On the opposite side of the room, the streetlight's distant luminance coating the window shade has an eerie unwelcome glow. None of these previously familiar objects have any familiarity now. What's worse, I cannot remember or recognize myself. I sit up in bed—actually, I lurch in mild sleepy terror toward the vertical. There's a demon here, one of the unnamed ones, the demon of erasure and forgetting. I can't manage my way through this feeling because my mind isn't working, and because it, the flesh in which I'm housed, hasn't yet become me.

Charlie wakes to a strange world, having lost who he is, for a moment, until his wife stabilizes him into self-recognition by putting her hand on his back. He gets up, goes to the study, and then descends the stairs, at the bottom of which he passes a large looking glass that does not reflect. The reason Charlie gives for this is that the mirror is so old, but we have to wonder, too, if he has somehow gotten to the other side of the glass. He goes out into the Ann Arbor night and wanders through his neighborhood, which is infested by gypsy moths. Out on the street, the traffic light blinks red in both directions—collapsing space, or perhaps slightly forbidding all things. He trespasses in the football stadium—where no one is supposed to be. Down below, right on the fifty-yard line, is a young couple, in the throes of sex.

Charlie finally comes across a man sitting on a park bench; this is his neighbor Bradley—who looks like a toad. Bradley has a dog that is disconcertingly also named Bradley. Bradley, the man, tells Charlie that it is illegal for anyone to be out so late in Ann Arbor without a dog.

All these temporal and psychic perversities combine to put us in an unstable situation in which something, everything, is bound to happen. We, and the characters, have entered into the world of the story, which is clearly a different world from our own. Baxter's mirror on the stairs in the opening of *The Feast of Love* reminds us of that other great mirror, Alice's looking glass. In both *Alice in Wonderland* and *Through the Looking Glass*, Alice is consumed by an intense drowsiness, and she tumbles, in the first case, and steps, in the second, into the other world.

Although the state between sleeping and waking occurs in my examples above, there are other ways to enter the entrancements of reading or writing. Day-dreaming works well, as does a certain sort of focused, almost electrically fierce, attention. In fact, anything that leads to a diminution of the actual world around us, anything that helps us to veer away from the indicative mood, can work here.

Descents or Passages to the Other World

When we enter the altered state of consciousness to which fiction carries us, we lose ourselves. We lose our sense of time. Language becomes strange or altered. We replace our loved ones—our lost, our missing, our absent, our longed for—with new people, sudden strangers who are at once accessible and beguiling. Though we cannot quite reach out and touch these strangers, they touch us, move us, break our hearts or heal them, and cause us to lurch into laughter and weeping. They make us know in ways we have never known before, and take us beyond human limits, beyond the borders of the mind.

In what follows I will try to show that when we achieve this altered consciousness, we are echoing the descent of the classical hero to the Underworld, also called the world below, the nether world, the world beyond. This is generally the Land of the Dead, the realm that the ancient Greeks called after the god Hades. I should emphasize that this is the case—our echoing of the heroic descent, I mean—whether or not the fiction we are reading has anything to do with the Land of the Dead. The works that I discuss here do contain descents to one sort of underworld or another, but my general claim about what happens to us as readers has to do with the nature of literary entrancement and not with the contents of any particular narrative.

It might be a good thing to clarify the idea of descent, often termed *katabasis* when the destination is the Land of the Dead. I should note that I am primarily interested in the living who go down to this region to visit, not the deceased who go down to stay for good. What I am calling *descent* has nothing to do with altitude relative to ground level. In cultures where the dead are buried underground, their territory is obviously thought of as being down below, but the geography of the Land of the Dead does not lend itself to rational mapping, and the journey can be one of infinite horizontal, rather than finite vertical, distance. It is as though having gone so far—to the edge of the world—we get to where space has buckled and "far" has been folded onto "under." Thus, the ancient Mesopotamian hero Gilgamesh, whom we will follow in a while, approaches the world beyond by traveling through the mountains where Day and Night live, outracing the sun, and finally crossing the River of Death. Odysseus, too, travels horizontally, crossing over the River of the Ocean to get to the edge of the world. Once he is there, the shades of the dead come up to visit him at ground level. The poet Orpheus makes the steep descent all the way down to the throne room of the King and Queen of the Dead, while the Greek poet and philosopher, Parmenides, travels, as we shall see when we look at his poem, "as far as longing can reach," before crossing the chasm of the underworld.

To some extent, this distinction between travelling across or down is beside the point: what really matters is the act of conversing with the inhabitants of the Other World once one gets there. Lewis Carroll's Alice tumbles *down* the rabbit hole in one book, steps *through* the looking glass in the other, but the worlds on the other side are functionally the same: totally Other.

Thus the world of the dead, or any of the other worlds, although often sharply demarcated and protected from the land of the living, can be in any direction, and one can descend, cross over, or even ascend to it.

Sometimes a true descent into death, a real dying as well as a brief heroic visit, is necessary before a final ascent to and permanent residence in Mount Olympus (in the case of Heracles) or Heaven (in the case of Christ).

❧ ☙

An important aspect of most heroic voyages to the land of the dead, or the land at the edge of the world, is the presence of one or many guides. The territory of the dead is so well protected from the living that getting there should not be undertaken alone if one plans to come back alive. In some cases there are multiple guides, with the first one helping with physical transport, acting as ferryman or charioteer, or simply giving directions to the limit of the dangerous realm.

At that point someone new—immortal, seer, or Goddess—takes over, directing the spiritual transformation.

The goals of such voyages or descents may be physical or material: to bring back a beloved, or a monster, or treasure.

In some ways, Christ's descent to Hell is atypical. For one thing, unlike the other heroes who make this trip, he has nothing to lose: his descent occurs after the crucifixion and before he goes up to Heaven, so he is already dead. For another, he does not go down for the usual reasons: to retrieve ideas, treasure, or a lover. Instead he descends to Hell in order to liberate a whole population of Old Testament patriarchs, matriarchs, and prophets, starting with Adam and Eve, Abraham, Sarah, Isaac, Solomon, David, and others. These righteous figures did not deserve to be in Hell, but because they died before Christ lived, they could not achieve Christian salvation in the normal way, by accepting him as their savior.

When Christ goes to the Underworld, he sacks and plunders the place, he smashes the gates. He *harrows* it. Two separate etymological strands go into the verb *to harrow*. The first one seems to have more to do with farming, coming from the Middle English word for *rake*; it means: to draw a harrow over; to break up, crush, or pulverize with a harrow; to plough; to tear, lacerate, wound (physically); to lacerate or wound the feelings of; to vex, pain, or distress greatly; to disturb (obsolete); or to castrate (obsolete). The other form of *harrow* is from the Old English word for *harry*, which means: to rob, spoil, or plunder.

When Christ harrows Hell, I think both senses of the word are in play, though most often it is used in the sense of plunder—which is rather odd, for if Hell is to be a form of eternal prison, you would think that its perimeter should remain intact. Still, Christ smashes the doors, as we shall see, to smithereens.

• •

Sometimes the goal of katabasis or a voyage to the other world is to find knowledge, religious instruction, wisdom, or revelation. Most of the ancient stories we will follow in Chapter II have to do with these abstract quests. We will also look at Katherine Mansfield's brilliant story, "The Garden Party," which illustrates a descent to the Land of the Dead in a more modern setting.

Katabasis is reminiscent of pearl diving: one explores the depths of a perilous medium to the limit of one's breath and life in order to bring back a rare and valuable treasure that could be found nowhere else.

Most descents, of course, are one way, made by the dead themselves. The Greek god Hermes, the Messenger, is the one who acts as psychopomp, or conductor of souls, and he is the only one of the Olympian gods who regularly goes back and forth between our world, the underworld, and the heavens. The later, post-Homeric attributes of Hermes as the god not only of literature but also of interpretation (called *hermeneutics*, "the Hermes art") may have sprung from his frequent proximity to the souls of the dead, whose speech is described in Book 24 of the *Odyssey* as being like the twittering of bats.[3]

☙ ❧

Related to all these journeys is the trip to *Nighttown*. James Joyce uses this term in *Ulysses* to refer to the red light district of Dublin, but I am redefining it here to mean the urban, living equivalent of the Land of the Dead. Certainly even my conception of Nighttown retains its erotic connotations, for Hades and Eros have been linked since antiquity, but I do not use it simply to refer to a tawdry part of town. Rather it has elements of the spiritual, of the uncanny, of Death and all the chthonic deities; sometimes it has hallucinatory qualities as well. Descent, or crossing over infinite distances, may not be necessary to get there; sometimes we find Nighttown in the soft underbelly of our own hometown. Nighttime is not always a prerequisite for Nighttown, and one of the most intense and beautiful fictional accounts of this place, which I will discuss in Chapter IV, is Eudora Welty's long short story "Music from Spain," where Nighttown is located in San Francisco, in broad daylight.

How do we know if we ourselves, or a set of fictional characters, are in our world or in some other world such as Nighttown, Hades, Wonderland, or Through the Looking Glass? The most important marker for these other worlds is a blurring of boundaries between things that are usually kept distinct, such as kingdoms, species, or genders. Thus man can be part beast, part god, or part woman. The invisible becomes visible. Time gets distorted, or turns out to be measured by odd clocks. Language can become strange or unintelligible, needing translation.

In general, our world consists of the known and the normal, while Hades, Nighttown, and other examples of the Other World are full of the unknown and the abnormal. For example, some of the indicators that Baxter's *The Feast of Love* takes place in the other world of what I call Nighttown include the narrator's waking when he should be sleeping, mirrors that refuse to reflect, the strange confluence of the dog Bradley with the man Bradley, and the human Bradley's own resemblance to a frog. In a final infringement of classical literary boundaries, it is the character of this frog-like man, Bradley, who tells Charlie Baxter, the narrator and author, how to write his book.

This last transgression—that of a character instructing his author—deserves comment, for I want to try to clarify the distinction between the *representation* of otherworldliness in the final text and the *experience* of otherworldliness in the act of writing itself.

At times, while we are writing, our characters take it into their heads to dictate to us. When this happens, the process can feel more like recording than inventing, more like pulling back the veil than like making things up from whole cloth. Who is the teller? we wonder, who the listener? Who is in control here? How can there be such instantaneous communication between beings in this world—me at my desk, the black cat who occasionally visits me—and beings of the other world of my fabrication? Swatches of the act of writing seem to take place otherwhere.

But I should stress that the made object—whether poem or story—will rarely have the other world as its setting. Even more rarely does the narrative reveal the otherworldliness of the act that went into its making. Baxter's novel, however, does both. With its porous boundaries between things that are usually kept distinct, *The Feast of Love* takes place in the other world of Nighttown while it also reflects, by means of characters who lecture the author, the uncanniness of the experience of writing.

❧ ❧

There is an important practice connected to artistic creation as well as to entry into other worlds; this is the ancient ritual of *incubation.* This practice was most famously linked with the cult of Asclepius, healer-turned-god, whose main temple was at Epidaurus.

Those seeking relief from malady or affliction would come to the temple in order to sleep in the sanctuary. This process would either cure them directly or it would lead to a dream or vision that would indicate, perhaps with priestly interpretation, the proper treatment.

Although most well known in its connection to Asclepius, incubation was also practiced by priests of Apollo, called healer-prophets, *iatromantoi*. The classical scholar Peter Kingsley makes the controversial argument that the Apollo worshipped by these priests was associated with healing, with snakes, with death and the underworld, and with caves where the incubation rites took place. Kingsley says that this Apollo was a god of darkness with very different attributes than the Apollo we associate with clarity and light.[4]

Incubation as performed by the healer-prophets of this dark Apollo required prolonged stillness and silence, and was often performed in caves, because of the lack of sensory stimulation. This appears to have been a meditative practice, more than a medical one, with the goal being the dream or vision itself, rather than a path to a remedy for a physical ailment. Most people wanting to incubate required a teacher to guide them in their practice, to watch over them during the trance state, and to keep them from being eaten by bears or other beasts.[5]

Figure 6. *Pythagoras Emerging from the Underworld* **by Salvator Rosa (see Plate II).**

In an oil painting by Salvator Rosa, now in the Kimbell Art Museum, we see the hooded figure of the ancient Greek mathematician Pythagoras just emerging from his incubation in a cave, greeted by a throng of enthusiastic and worshipful followers (Figure 6). It looks as though they have been waiting for some time. Ancient sources made the strange claim that when he descended to Hades, Pythagoras

> saw the soul of Hesiod bound fast to a brazen pillar and gibbering, and the soul of Homer hung on a tree with serpents writhing about it, this being their punishment for what they said about the gods.[6]

It is not clear whether this descent was part of a routine incubation, or a different sort of voyage. It is also unclear whether the Hellish punishments were because the poets had written lies—saying that the gods had very human lusts and passions and vengeful rages—or because they had revealed esoteric knowledge of divine truths. Certain types of divine knowledge have always been forbidden, and always connected with danger; some of these kinds of forbidden looking we will examine in Chapter III.

Incubation can be thought of as a magical-religious analogue of a hero's voyage to the other world; it feels like an echo or re-enactment of such descents and voyages, complete with spirit guide and attainment of knowledge, wisdom, or revelation.

Incubation is closely related to that beloved and necessary activity of the writer: day-dreaming. While true incubation in lonely caverns may be more prolonged and more extreme—and its trance states deeper—than day-dreaming, both of these forms of stillness allow us to put the world away for a while in order to come up with dreams and visions, poems, laws, solutions. Writing is, aside from mathematics, probably the only pursuit where day-dreaming is so sought after and cherished. Musing, we call it, for that is when we let down our rational guard, turn off our censors for what is reasonable, and open ourselves to the Muses.

Lascaux

Before we follow the hero to the Land of the Dead, I would like to recount a contemporary descent, which was my first exposure to the subterranean mysteries, and which led to my fascination with such descents.

The year is 1988. The place is the prehistoric cave at Lascaux in Southern France.

Though the hills are not far away, the land nearby is flat and we can't see any cave openings as we stand by the fence surrounding the site. The *Laissez-passer*, our permission slip from the Ministry of Culture, has all three of us inscribed—my husband Barry, our philosopher friend Eva Brann, and me, and it has directed us to this unmarked gate in the middle of nowhere in the Dordogne valley in southern France. The attached instructions say that we will have to undergo disinfection procedures before our descent.

We wait, in the soft rain, wondering if it is the right fence, the right rusted gate. There are no signs. We wonder if we're in the right place, but we don't mention our doubts to each other. Instead, Eva asks us if we would recognize the people who painted these caves 17,000 years ago. "Do you think we would know them as *us*?" she asks.

A man dressed in blue denim ambles toward us from the other side, jingling his keys. Without greeting us, he unlocks the gate and silently leads us to a stone hut. There he examines our papers with ecstatic grimness, as though he's hoping our documents will not be in order so he can send us away again. But he changes into an army fatigue jacket and tells us to leave our sacks on the table. I proudly show him our new flashlights but he orders us to leave them behind as well. "*C'est moi qui prendra la lampe,*" he says. "I'll be the one taking the lamp. I'm the guide, I'll be showing *you* the cave. It's better like that, no?"

Out of the hut and down some cement steps into a bunker with a heavy steel door. Eva rephrases her question, "What if one of them were coming out, one of the cave painters? Would we recognize him?"

We don't have time to answer her. Our guide motions us inside and steps into a footbath demonstrating how deeply we should shake our shoes in the disinfectant. This is meant to kill any hitchhiking microorganisms, as molds and bacteria brought in by travelers have tended to attach to the cave walls and eat the paintings to such an extent that Lascaux has been off limits to the public for decades. One has to apply a year in advance to get in. The room smells of the sweetly vicious odor of formaldehyde.

Steel doors clang shut behind us and lock with a thud. Our guide opens a smaller door on the opposite side and herds us down steep iron-grill stairways, barely lighting our way with the narrow beam of his flashlight. Then he positions us by the cave wall and turns off his light, saying that it is time for us to become dark adapted. The blackness is thick and velvety and sudden. Somewhere behind us water drips to a pool.

As we stand there bewildered in the darkness, the man talks to us, telling us that the cave was discovered in September 1940 by four schoolboys. We want to ask him which of the two discovery legends is correct. The one in which four boys out hunting were looking for their dog that had fallen down a hole made by an uprooted tree . . . or the other one, where the four boys were following the instructions of an old woman who told them of a cave with what she called "medieval drawings." But when we start to ask him, he admonishes us to keep quiet until we emerge. Our time below is limited, he says, and he will be doing all the talking. We stand in the darkness, obedient children, unnerved.

Finally he turns on his light again and points out two types of cave wall: the smooth part, generally used for incised drawings, and the rough part, encrusted with microcrystals of calcium carbonate. This rough wall is the canvas used for the paintings of Lascaux, and it is part of the reason they have lasted for 17,000 years. The hard

Figure 7. "The Shaman" of Lascaux.

needle-like projections of the tiny crystals capture and hold the pigments. Like a sort of mineralized velvet, this surface doesn't allow any erasures—what was drawn is what you see.

Down a narrow passage to the Chamber of the Bulls. Five or six bulls follow each other, facing into the depths of the cavern, with the first one we meet, the one closest to the exterior, having a somewhat different shape and two long horns with blobs at the ends. The whole beast looks a bit saggy and off somehow. Our guide tells us to keep this one in mind. He will discuss it later (see Figure 7).

Except for this first animal, throughout the cave the sureness of line and form is astounding, as is the freshness of the painting, the intensity of the pigments. The cavern's own projections and recesses are used to play into the volumes of the painted forms. Sometimes a hollow in the rock is reverse-shaded to give the optical illusion of the bulging side of a cow. Some animals extend around corners in the rock, so that while painting one end of the beast the artist could not have seen the other end, and yet a nubbin of rock at the head is used for the eye, a spine of rock toward the tail is used for the line of the haunch, and the whole animal is in perfect proportion.

Five reindeer swim across a river, their heads raised to keep their muzzles above the water, their antlers graceful and precise.

A horse falls off a cliff, the cliff being a natural outcropping of the rock. The horse is upside down, legs stiff in the air, ears back in fear, the mouth open in a sharp whinny.

Our guide turns off his light again and has us walk downhill along the slippery path. Then he tells us to turn around as he illuminates a larger cavern.

Everywhere, on every surface of the folded walls and ceilings, there are cows, bulls, goats, reindeer, horses, bison—dancing, billowing like clouds. We are in "The Sistine Chapel" of Lascaux (see Figure 8).

By this time, all three of us are weeping, and seeing that we are properly moved, our guide has warmed and gentled.

At last he leads us back to the first animal in the cave, back in the Chamber of the Bulls. There he shows us how, if we cover the top part of the muzzle of this being with one of our hands, we can suddenly see the whole figure as a bearded man, wearing a sagging animal skin and costume horns. Clearly, those long straight horns with round blobs on the ends belong to no known beast, and the eye really seems to be a human eye, not an animal one. Our guide has saved this figure for

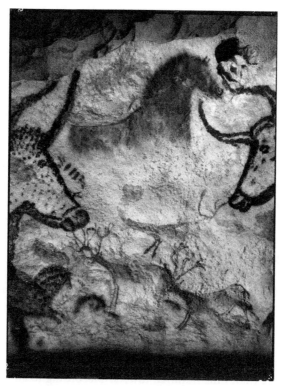

Figure 8. "The Sistine Chapel" of Lascaux.

the very end of his tour when we have seen enough other drawings to realize that the artists here are in total control of their craft, their art. So, there is a reason for the apparent clumsiness of the skin on this first bull: it is not his own skin but a ritual costume, worn by a shaman.

Our guide leads us up into daylight again. When I get my voice back, I ask him how long he has been taking people through the cave.

"*Je prends ma retraite dans deux ans*," he says. "I'll retire in two years. I've been at it for forty seven years."

"Ah, then you're the master of the cave," I say glibly.

"*Ah, non non non*," he corrects, gesturing back towards the cave entrance. "*They* were the masters."

Late that afternoon, back at our hotel, I am still shaken by what I have seen, though I don't yet realize that this exploration will lead to my long-lasting obsession with the world below. Leafing through a small booklet on Lascaux, I notice a recent picture of our guide, tending some hygrometric machinery having to do with the preservation of the cave. His name is given as Jacques Marsal. On page one of the same booklet is a picture of the four boys who discovered the cave in 1940: Jacques Marsal is one of them.

Gilgamesh

Stories of crossing into the Other World or the Land of the Dead are marvelously old. The earliest accounts we have are from Mesopotamia, the land between the Tigris and Euphrates Rivers, corresponding to parts of what is now Iraq, as well as parts of Syria, Turkey, and Iran. The most intense and moving of these tales is the *Epic of Gilgamesh*. The earliest pieces we have of this story, in Sumerian, are around four thousand years old, but the "standard" version dates to 1300–1000 BC. This was redacted in the Semitic language of Akkadian, by Sin-liqe-unninni—a professional exorcist—whose name means "Oh Moon God, Accept my Prayer."

Gilgamesh was King of Uruk, the biblical Erech, in 2800 BC. The walls of Uruk, which he is said to have built, were nearly six miles long and had more than nine hundred towers.[7] According to an ancient list of Mesopotamian kings, his father was a phantom and his mother was the goddess Ninsun, a deified wild cow.[8] While alive, Gilgamesh was two-thirds god, one-third human, an impossible ratio in any scheme of human genetics, indicating that we are not dealing with the usual mortal realm.

In the epic that bears his name, Gilgamesh becomes terrified by the idea of mortality when his beloved companion, Enkidu, dies. He then tries valiantly and unsuccessfully to avoid such a fate for himself. Ironically, a few hundred years after his own mortal death, the real Gilgamesh did achieve immortality, becoming a god who was worshipped as a judge of the Underworld.

Here is the story: As a young king, Gilgamesh terrorizes his subjects with riotous energies and adolescent lusts, often invoking *droit du seigneur*. Finally the citizens of Uruk complain to the gods, and the gods decide to create a counterpart for him, a companion who can absorb his energies. This created counterpart is the feral Enkidu, a man raised in the wild by animals, a man who still drinks at watering holes by lowering his head to gulp the water. A prostitute is sent to lure Enkidu from the wild and tame him. She couples with him for six days and seven nights and teaches him the ways of civilized man. Now all the wild animals flee from him.

Learning more of civilized ways from shepherds, Enkidu hears of Gilgamesh's bad behavior and goes to Uruk to confront him. They fight viciously, become best friends, and go off together to seek glory. After killing the ogre Humbaba, who is guardian of the Cedar Forest, they steal the giant cedar to make the great doors of a temple at home in Uruk. Attracted by all these heroics, the goddess Ishtar offers herself to Gilgamesh. When Gilgamesh refuses her advances and further insults her, Ishtar gets her father to send the Bull of Heaven (Taurus) down to earth to kill him. Gilgamesh and Enkidu sacrilegiously kill the bull. After a huge celebratory feast, Enkidu has nightmares, falls sick, and dies. Perhaps this is the first historical instance of food poisoning from a public gathering. Gilgamesh weeps over Enkidu's body for six days and seven nights and does not give up the corpse for burial until finally a maggot drops from Enkidu's nostril. Following a vast royal funeral, Gilgamesh leaves Uruk and goes into the wild:

> "I shall die, and shall I not then be as Enkidu?
>   Sorrow has entered my heart.
> "I am afraid of death, so I wander the wild,
>   to find Uta-napishti. . . ."
>
> [IX:1]

This Uta-napishti is the Mesopotamian Noah figure. Survivor of the Great Flood, he lives with his wife at the end of the world, the only humans to achieve immortality. Gilgamesh hopes to learn from him how to outwit death. During his travels, the young hero kills lions, eats their flesh, and dresses in their skins, becoming progressively shaggier in looks and behavior, turning as feral as his beloved Enkidu had been, before the prostitute civilized him.

The sun god, Shamash, warns Gilgamesh that this mission to go to the end of the world is futile, but the youth scorns divine advice and continues until he comes to the twin mountains whose peaks "support the fabric of heaven" and whose base "reaches down to the Netherworld." The entrance between these mountains is where the sun rises and sets, and it is guarded by scorpion-men "whose glance was death, whose radiance was fearful."

The scorpion-men ask Gilgamesh a set of crucial questions, which I will call Traveler's Questions (I will talk more about these questions in the next section):

> "How did you come here, such a far road?
> How did you get here, to be in my presence?
> How did you cross the seas, whose passage is perilous?
> . . . let me learn of your journey!"

> [IX:55]

Gilgamesh explains his quest, the scorpion-men let him pass, and he enters the intense darkness within the mountains where he races for twelve "double hours" and succeeds in beating the sun to the other side. He emerges into a jeweled garden:

> . . . the trees of the gods.
> A carnelian tree was in fruit,
> hung with grapes, lovely to look on.
>
> A lapis lazuli tree bore foliage,
> in full fruit and gorgeous to gaze on.
> (. . .) instead of thorns and briars there grew stone vials.
> He touched a carob: it was *abashmu*-stone,
> agate and haematite. . . .

> [IX:190]

This may well be the first written account we have of undersea coral gardens, or skin diving exploration. Again we have the resonance between entering the world beyond and diving for pearls.

Soon Gilgamesh comes to a goddess who keeps a tavern by the sea. At first she mistakes him for a hunter or wild man and will not let him in, but he threatens her, proclaiming that he is the king-hero Gilgamesh. Disbelieving, she retorts,

> "Why are your cheeks so hollow, your face so sunken,
> your mood so wretched, your visage so wasted?
> Why in your heart does sorrow reside,
> and your face resemble one come from afar?
> Why are your features burnt by frost and by sunshine,
> and why do you wander the wild in lion's garb?"

> [X:35]

Gilgamesh tells of his intense grief, equaled only by his fear of death. When he asks for directions, this tavern-keeper goddess is adamant that only the sun god Shamash can cross the ocean, for

> "The crossing is perilous, its way full of hazard,
> and midway lie the Waters of Death, blocking the
> passage forward. . . ."

[x:84]

But she relents and tells how to find the ferryman, Ur-shanabi, who can take him across the Waters of Death. Inexplicably, when Gilgamesh finds the ferry, he destroys the stone men (or idols) who usually crew (or protect) the boat. The ferryman instructs him to go to the forest and cut three hundred punting poles. This seems very strange until we hear the warning when they launch: "Set to, O Gilgamesh! Take the first punting pole! / Let your hand not touch the Waters of Death, lest you wither it" [x:175].

In punting, one pushes a long pole into the river bed, propelling the boat forward until one can just grasp near the top of the pole; then

Figure 9. River scene in Guiné-Bissau.

one lifts the pole, much of which is now wet, sliding it through one's hands until it is in position to be thrust down once more into the river bed. It is while repositioning the pole through his hands that our hero would come in contact with the deadly withering waters, so each pole must be discarded before it is pulled up again. This is why Gilgamesh will need 300 of them. In many parts of the world such long punting poles are still used to propel and steer long narrow boats (Figure 9).[9]

The Land of the Dead is extremely well protected from incursions by the living, almost as though contamination might occur. This is not a realm for the idle tourist. One needs multiple guides. Instructions must be followed to avoid fatal consequences. It is both too hard and too easy to come to the world of the dead. Too hard to find one's way there while alive, I mean, and too easy to die in the attempt.

When Gilgamesh finally lands on the other side of the Waters of Death, he tells his story to the human immortal, Uta-napishti. This wise being takes one look at the filthy lion-skin clad youth and gives him a lecture about the difference between kings and fools. He instructs Gilgamesh to start acting like a proper king, reminding him that one of the duties of a ruler is to provide for the temples of the gods and goddesses. Then he gives Gilgamesh a strikingly beautiful discourse on death:

> "(. . .) Man is snapped off like a reed in a canebrake!
> The comely young man, the pretty young woman—
> all too soon in their prime Death abducts them!
>
> "No one at all sees Death,
>   no one at all sees the face of Death,
> no one at all hears the voice of Death,
>   Death so savage, who hacks men down.
>
> "Ever do we build our households,
>   ever do we make our nests,
> ever do brothers divide their inheritance,
>   ever do feuds arise in the land.
>
> "Ever the river has risen and brought us the flood,
>   the mayfly floating on the water.
> On the face of the sun its countenance gazes,
>   then all of a sudden nothing is there.

"The abducted and the dead, how alike is their lot!
But never was drawn the likeness of Death,
never in the land did the dead greet a man."

[x:318]

Uta-napishti gives Gilgamesh a test to see if he is really ready for immortality: as a model of avoiding death forever, can he do without sleep—that analogue of death—for a week?

Gilgamesh fails miserably, falling into a narcoleptic slumber as soon as he hunkers down on his haunches. Uta-napishti tells his wife to bake a loaf of bread each day and put it beside the sleeping hero, so that there will exist incontrovertible proof: the progressive staleness and moldiness of the daily loaves will show the length of time he has slept. When he wakes, a week later, Gilgamesh understands that he has no hope of avoiding Death. He howls:

"O Uta-napishti, what should I do and where should I go?
A thief has taken hold of my flesh!
For there in my bed-chamber Death does abide,
and wherever I turn, there too will be Death."

[XI:246]

Uta-napishti commands the ferryman, Ur-shanabi, to wash Gilgamesh until he is clean, then to dress him in robes appropriate to his royal status. As a consolation, Uta-napishti tells Gilgamesh where to dive for a secret plant-like coral that makes one young again. When Gilgamesh carelessly goes for a swim after retrieving the plant, a snake comes along and eats it, then sloughs off its skin in an act of renewal. Gilgamesh sits and weeps over the uselessness of all his toils. He is a wondrously weepy hero.

When he arrives home, however, Gilgamesh exults as he shows his ferryman his city of Uruk, with its solidity, order, and beauty. He recognizes that Uruk is a joy to gods and men, a lasting construction.

To the extent that his mission was to conquer Death, it was a complete failure. But it is as though the story knows, if Gilgamesh does not, that the true purpose of exploring the world beyond our own is a deeper understanding of mortality, and how human creations can be the paths to human and divine pleasure.

જ ક

Earlier I mentioned that some indications that we are dealing with the other world—whether the one below, across, or infinitely distant—are a blurring of categorical boundaries, a distortion of time, and a strangeness of language. In the story of Gilgamesh, although there are no problems of language, several boundaries have become indistinct. The guardians of the entrance to the twin peaks of the mountain of the sun are chimerical scorpion-men showing the loss of demarcation between man and animal. The line between the mineral and vegetable kingdoms is erased as well, for on the far side of that same mountain is the jeweled garden with flowers made of precious stones. And finally, the distinction between mortal and immortal is erased by the human survivors of the flood, Uta-napishti and his wife, who are immortal.

There is also something very strange happening with time, as though the whole tale has an obsession with, or hypersensitivity to, time. When Gilgamesh goes through the mountains where the sun both rises and sets, he races the sun for twenty-four hours, and each of the twelve "double hours" is chronicled separately but with highly repetitive wording:

> At one double hour
> the darkness was dense, and light there was none,
> it did not allow him to see behind him.
> At two double hours
> the darkness was dense, and light there was none,
> it did not allow him to see behind him.

[IX:144]

So it continues for many lines, with very minor variations through the hours of the single day. The poetry here feels like a counting song, and perhaps has the same undercurrent of religious meaning that many counting songs have. Gilgamesh wins the race, that is, he comes out through the sun's own mountain first, getting to the morning of the next day before the sun god, traveling faster than the light. The incantatory quality of the text here feels as though the tale is taking a recess, to show us an entrancement taking place: the story is entrancing itself and its listeners, and reminds us of how we, too, fall under enchantment when we read.

Later, further preoccupations with time occur when time is measured in loaves of bread, as proof of the sun god's diurnal passing, while Gilgamesh demonstrates his inescapable mortality by sleeping. Gilgamesh's competition with the great celestial time-piece shows itself not only in the obsessive chronicling of the twenty-four-hour race, but also in this strange chronometry of loaves.

∾ ∾

There is something thrillingly unadorned about Gilgamesh and his quest. Fear of death, or coming to terms with mortality, may underlie every hero's mission but usually it is disguised, camouflaged, or transfigured into something else. Gilgamesh is so brash and unveiled; there is no protective metaphoric covering for the brilliance of his purpose or the starkness of his fear. "O, Uta-napishti . . . a thief has taken hold of my flesh."

Incursion into the other world has left our hero more peaceful and more human. This is partly because of the way his guides instruct him, and partly because such voyages—if they are profound enough, if we pay close and deep attention—always make us wiser.

Odysseus

Our next account of a visit to the other world also takes place within the context of an immense and eventful voyage. At the end of the Trojan War, Odysseus sets out for home, but by the time he finally gets back to Ithaca, he has stayed away for a decade and has succeeded in losing all of the men under his care.

In the middle of *The Odyssey*, book XI, Odysseus sails to the Land of the Dead. For this journey he has two guides. Circe, the witch, tells him how to get there, and also how to behave, so that his real guide, Teiresias, the blind Theban seer, will appear and tell him everything he needs to know. Teiresias will advise Odysseus how to get home, and he will also teach him how to deal with the dead, how to make them tell their stories to him in an intelligible manner, one at a time.

According to Circe's travel instructions, quoted here in Robert Fitzgerald's translation, Odysseus must sail before the North wind to

> the bourne of Ocean,
> Persephone's deserted strand and grove,
> dusky with poplars and the drooping willow.
> Run through the tide-rip, bring your ship to shore,
> land there, and find the crumbling homes of Death.
> Here, toward the Sorrowing Water, run the streams
> of wailing, out of Styx, and quenchless Burning
> torrents that join in thunder at the Rock.[10]

By the grove of poplars and willows, by the confluence of the terrifying rivers coming from the Underworld, Odysseus follows Circe's instructions, digging a hole as broad and as deep as his forearm. He pours in milk and honey, wine, then water, and he sprinkles in barley; he promises future sacrifices to all the dead. Then he slashes the throats of a black ram and ewe, letting their blood flow into the votive pit. When the dead get wind of this blood, their hungry souls come flowing up from Erebos, the Underworld. This indicates that Odysseus has not made a physical descent; the part of the Domain of Hades that he is visiting appears to be roughly at sea-level, and Erebos is located underneath.

Two dreadful surprises appear among the approaching souls: Elpenor, one of his young shipmates, has preceded him to the Land of the Dead, having died of a broken neck after falling from Circe's roof. The next recognizable shade is also a shock: Odysseus's mother, Anticleia, who was still alive when he left for Troy. Odysseus, though he weeps to see her there, will not let her near the blood until he first talks to Teiresias.

In a red-figured calyx-krater in the Bibliothèque Nationale de France, we see Odysseus, with his sword, and with the slaughtered ram and ewe at his feet (Figure 10). The grey-haired face of the ghost of Teiresias, pokes up from the ground, at the level of the votive pit. Eurylochos, stands to the left, wearing a brimless cap and short cloak. He clearly does not listen to the seer, for he will be drowned after carrying away the cattle of the Sun God, an act advised against by Teiresias.

Teiresias addresses Odysseus as "son of Laertes and the gods of old," for Zeus is Odysseus's paternal great-great-grandfather. It is often

Figure 10. Odysseus consulting Teiresias in Hades by the Dolon painter.

the case that the hero who successfully crosses to the other world is of divine lineage. Teiresias goes on:

> Odysseus, master of land ways and sea ways,
> why leave the blazing sun, O man of woe,
> to see the cold dead and the joyless region?
> Stand clear, put up your sword;
> let me but taste of blood, I shall speak true.

[XI:107]

Like all the dead, Teiresias thirsts for life fluids—we will talk more of this when we discuss the Evil Eye, in Chapter III. But even before he mentions his desire to taste the fresh dark blood, Teiresias asks one of what I call the Traveler's Questions:

- ❧ Where have you come from?

- ❧ How did you get here?

- ❧ Why have you come?

These Traveler's Questions ask for the physical and emotional coordinates of the stranger, and the answers will help to triangulate him according to his origins, means, and intentions. The questions are important not only because of their universality, but also because they remind us that the descending hero often gives, as well as receives, information. In the case of Odysseus, many of the dead also ask for specific news, usually about the well-being of their own family members in the world of the living.

Odysseus's mother, Anticleia, asks two of the questions, first the *how*:

> "Child
> how could you cross alive into this gloom
> at the world's end?"

And then the *whence*:

> "Say, now,
> is it from Troy, still wandering, after years,
> that you come here with ship and company?
> Have you not gone at all to Ithaka?
> Have you not seen your lady in your hall?"
>
> [XI:172]

Even though Teiresias is a seer whose prophetic counsel is the reason that Odysseus has come to Hades, even he still asks the *why*, "Why leave the blazing sun, O man of woe, to see the cold dead and the joyless region?"

ॐ ॐ

As writers, and as readers, we are constantly in the sway of the Traveler's Questions. These are questions we ask of our characters while invoking or inventing or revealing them. More importantly, they are what the text we are writing, or reading, keeps asking of us. For our texts interrogate us when we have crossed into the world of the imagination, until we can figure out why we have come, where we have come from, and how we have managed the passage.

❧ ❧

The world of *The Odyssey* is largely a world of women—goddesses and nymphs and mortals—and even when Odysseus visits Hades, after he talks to his mother, it is the famous women who appear to him, many of them former consorts of Zeus. "Here was great loveliness of ghosts!" Odysseus says, and he recounts the stories of at least a dozen of the women before telling of seeing Agamemnon, Achilles, and Ajax. When Odysseus flatters Achilles, surmising that his power continues to be royal even in Hades, "among the dead men's shades," Achilles bursts out:

> Let me hear no smooth talk
> of death from you, Odysseus, light of councils.
> Better, I say, to break sod as a farm hand
> for some poor country man, on iron rations,
> than lord it over all the exhausted dead.

> [XI:581]

The scene shifts away from the votive trench full of blood, and Odysseus now catches glimpses of the mythological dead within their own mythological landscapes: Minos, the king of Crete and son of Zeus, "dealing out justice among ghostly pleaders"; Orion with his bronze studded club; Tityos, who had once raped Zeus's mistress Leto, and now has vultures eating his liver; Tantalus up to his chin in a cool pond whose waters disappear to dry mud whenever he bends to drink; and Sisyphus pushing his perpetual boulder up the hill.

Finally Heracles comes and addresses Odysseus, and what is most odd is that Heracles is a ghost of a ghost. Following his real death (not his first descent to get the dog Cerberus as part of his twelve labors), he, like Christ, has ascended to the heavens where he has become one of the gods. Heracles seems to be asking a perfunctory form of the Traveler's Questions, but quickly turns the conversation to himself:

> Son of Laertes and the gods of old,
> Odysseus master mariner and soldier,
> under a cloud, you too? Destined to grinding
> labors like my own in the sunny world?

> [XI:734]

What is it that Odysseus brings up from the world of the dead? He learns that fierce lesson from Achilles that it is better to be the most servile laborer among the living than king of all the dead. Teiresias instructs Odysseus not to raid the cattle of the sun god, else he will lose ship and crew (both of which come to pass) and warns him about the treasonous and insolent suitors who are occupying his house and wooing his wife. Teiresias also instructs him in the ritual he is to perform when he gets home: he is to hike away from the sea carrying his wooden oar until he comes to men who have lived

> with meat unsalted, never known the sea,
> nor seen seagoing ships, with crimson bows,
> and oars that fledge light hulls for dipping flight.*

[XI:139]

There, where people mistake his oar for a farming implement, Odysseus is to plant it in a memorial mound and sacrifice to Poseidon. Only then can he go home and live in peace.

కఌ ఌ

Remarkably, as Eva Brann points out, Teiresias's instructions take up only one of the twenty-odd pages of Book XI of *The Odyssey*. All the rest is stories, the myths of the Greeks, and these are what Odysseus really brings back from the world of the dead. Brann calls Hades "the safe-depository of tales, the treasure house of myth." And these stories, she says, are his booty, his wise findings, his revelations:

> To Hades are consigned the stories that will become the common property of the Greeks. And from its depths they can be mined as is precious metal; they can be retrieved as from a place on the far side of mortal habitation; they can be brought back to life by a singer, a poet. The raw material of Musically crafted song comes to a poet by conversation with the souls of the dead, achieved by a sailing to the realm beyond and below.[11]

The world below as a repository of stories! This vision occurs in a stranger form in the novel *Sanatorium under the Sign of the Hourglass*,

---

* For me this one line, with the verb "fledge" proves the superiority of Robert Fitzgerald's translation.

by the Polish writer Bruno Schulz. For Schulz the underworld is vast; it is also as close to home as the nearest forest. In this subterranean region, words "return to their etymology, re-enter their depths and distant obscure roots." This process of words dissolving into their components is, Schulz says, to be taken literally: for him word roots and tree roots are totally entwined. He lures and invites us to penetrate among the roots of trees and the "smell of turf and tree rot," where, he claims,

> the interior is pulsating with light. It is, of course, the internal light of roots, a wandering phosphorescence, tiny veins of light marbling the darkness, and evanescent shimmer of nightmarish substances.[12]

Down here, in these "dark foundations," among the labyrinths of tree roots,

> There is a lot of movement and traffic, pulp and rot, tribes and generations, a brood of bibles and iliads multiplied a thousand times. Wanderings and tumult, the tangle and hubbub of history! . . . All the stories we have heard, and those we have never heard before but have been dreaming since childhood—here and nowhere else.[13]

Along with the deconstituted words, stories are there, waiting, forming a sort of fundamental warehouse or safe-deposit vault for the writer:

> Where would writers find their ideas, how would they muster the courage for invention, had they not been aware of these reserves, this frozen capital, these funds salted away in the underworld?[14]

Although Schulz is usually a wild and sensual writer, his description here is unusually abstract. Stories, etymologies, histories, bibles, and iliads are in the tumult down below, but no conversations, and no dead individuals to have them with. He argues that Spring is actually the resurrection of all these subterranean histories, and that is what makes that lovely season so rich and so sad. Thus, his view is a sort of distillation of archaeology—what we find when we begin to unearth the past—not only shards and buildings but histories, not only histories but stories, not only stories but words. The fundament of everything that counts.

The Homeric view of Hades, peopled with the shades of the dead, is more attractive to me. I need people down there; simple excavation of subterranean riches is not enough to supply the poet with raw materials. We have to converse with the dead as with the living. We have to face the Traveler's Questions—the Whence, How, Why—and then continue with talk that extends all the way from the roots of longing to the tendrils of desire.

Schulz's notion of the phosphorescence of roots is an important one. In the dark places, such glimmering can be what allows us to follow the path. Some words, for example, simply glow, and as we break them down to their earlier components, we follow their tracks until we get to what seem to be elementary glowing particles in the earliest and deepest realms of hypothesized proto-languages. Exactly what makes a word, a phrase, or a line of a poem glow is not easy to describe; it must depend on some energy given off as our mind resonates with the energy contained in the word form itself. The resulting illumination keeps our gaze fastened and causes the ferocity of our attention.

All those words, stories, images, and conversations with the shades of the dead occurring under the ground—I wonder if this could be another reason for using caves as the place for spiritual incubation. There we are closer to the roots of things, closer to history, to songs and stories, both known and unheard of. The cavernous proximity. "What a buzz of whispers," Schulz says. "What persistent purr of the earth! Continuous persuasions are throbbing in your ears."[15]

Parmenides

The ancient Greek poet and philosopher Parmenides goes to the Land of the Dead and comes back with ideas that he recounts in the form of a grand poem in three parts. The only part I will consider here is the Proem, or Introduction, in which he tells of his voyage to the underworld.

The Proem

The mares that carry me as far as longing can reach
rode on, once they had come and fetched me onto the legendary
road of the divinity that carries the man who knows
through the vast and dark unknown. And on I was carried
as the mares, aware just where to go, kept carrying me

straining at the chariot; and young women led the way.
And the axle in the hubs let out the sound of a pipe
blazing from the pressure of the two well-rounded wheels
at either side, as they rapidly led on: young women, girls,
daughters of the Sun who had left the mansions of Night
for the light and pushed back the veils from their faces
with their hands.
There are the gates of the pathways of Night and Day,
held fast in place between the lintel above and a threshold of stone;
and they reach up into the heavens, filled with gigantic doors.
And the keys—that now open, now lock—are held fast by
Justice: she who always demands exact returns. And with
soft seductive words the girls cunningly persuaded her to
push back immediately, just for them, the bar that bolts
the gates. And as the doors flew open, making the bronze
axles with their pegs and nails spin—now one, now the other—
in their pipes, they created a gaping chasm. Straight through and
on the girls held fast their course for the chariot and horses,
straight down the road.
And the goddess welcomed me kindly, and took
my right hand in hers and spoke these words as she addressed me:
"Welcome young man, partnered by immortal charioteers,
reaching our home with the mares that carry you. For it was
no hard fate that sent you traveling this road—so far away
from the beaten track of humans—but Rightness, and Justice.
And what's needed is for you to learn all things: both the unshaken
heart of persuasive Truth and the opinions of mortals,
in which there's nothing that can truthfully be trusted at all.
But even so, this too you will learn—how beliefs based on
appearance ought to be believable as they travel through
all there is."

This translation of the Proem is by the brilliant and controversial classical philosopher Peter Kingsley, who points out a number of strange things about it.[16] The first is that though Parmenides is thought to be the founder of Western Logic, in this voyage to the goddess who will teach him "all things," Parmenides does not say that his path proceeds from darkness into light. Rather it is the other way around; he is being taken from our world of light, through the vast and dark unknown, down to where Day and Night reside, back to "the mansions of Night" where the maidens have come from.

Kingsley argues that this passage into darkness becomes easier to understand if we note that Parmenides describes himself as "the man who knows," meaning one who is initiated into the mysteries, a seer. In Kingsley's view, these mysteries have to do with the incubation practices I mentioned earlier, and Parmenides was actually the founder of a succession of healer-prophet sects.

Like Odysseus, Parmenides is surrounded by females: mares pull his chariot; maidens guide the way; Justice opens the gates; and finally it is the goddess Persephone who greets and will teach him.[17]

As is often the case, there are two sorts of guides: the maidens and their horses, who physically transport the poet over the long road and down into the underworld, and Persephone, who takes over as spiritual guide, once they have arrived.[18]

Although the wording and the journey seem passive on the part of Parmenides—"the mares that carry me," "the divinity that carries the man," "and on I was carried," "the mares, aware just where to go, kept carrying me,"—it is important to remember that the only one who can travel this road alive is the "man who knows," that is, the initiate who has done all the necessary spiritual preparation beforehand.

I mentioned earlier that blurrings of boundaries and distortions of language or time often occur in accounts of the Other World. Parmenides travels to the domain of both Night and Day—a place where these crucial time markers coexist and thus would seem to lose their distinctive qualities—a region beyond time that echoes the mountain Gilgamesh traverses when he outraces the sun god.

The Traveler's Questions occur in an odd form in the Proem. That is, they appear by implication, for the Goddess gives the answers herself as though she does not even need to ask them. She knows how Parmenides has found his way—guided by maidens, carried by mares—and she knows that to get there, he has been carried "as far as longing can reach," which we know is infinitely far—"longing we say, because desire is full / of endless distances," as the contemporary poet Robert Hass says.[19] Persephone knows, too, how it is that Parmenides does not have to be dead in order to come to her domain—because the supreme conjunction of Rightness and Justice have allowed him to enter. And finally she knows, and possibly informs him, *why* he has come: to learn all things, both universal truths and the false opinions of men.

It would be such a comfort, to meet such a divinity who knows our Traveler's Questions, our queries of existence, and tells us the answers without our even having to formulate or utter the question.

But when we are acting as writers, we must be the ones who have this knowledge and who must provide this comfort for our characters—our creatures—and, by extension, for our readers.

҂ ҈

Peter Kingsley points out that there is something else going on in the Proem, which is the strange noise of the pipes or the whistling roar. Parmenides is an extremely careful poet, Kingsley says, and when he repeats words like "carry," he does it for incantatory effect. The poem is both describing and enacting a trance induction of the sort practiced in the ritual of incubation. Parmenides uses the Greek word *syrinx* for "pipe," and he repeats this word when he tells how the huge doors spin open, rotating in hollow tubes or *pipes*. The word *syrinx* is generally used either for a musical instrument or for the part of the instrument that makes the piping or whistling sound. This instance in the Proem is apparently the only time in the Greek language when *syrinx* is applied to the architectural hardware of doors, and it is supposed to give a sense of the sound that the doors make as well as recalling the hollow flute-like cylinder that surrounds both the axle of a wheel and the hinge pin of the door.

For Greeks, Kingsley says, this sound of piping and whistling was also the sound of snakes, which were sacred to Apollo. In Ancient Greek accounts of incubation, Kingsley argues, one of the signs that mark the point of entry into another world—into the trance state—is that one becomes aware of a rapid spinning movement; another sign is the vibration produced by a piping, whistling, hissing sound. This sound is also part of ancient exercises in breath control. And finally, it is an ancient—as well as contemporary—call for attention: Pssssst! or, perhaps, I think, even the call for silence: Ssshhh! Greek mystical texts, says Kingsley, explain that this hissing or piping sound, this sound of silence, is the sound of creation: the noise made by stars and planets as they coil and spin. I think there is also a bit of terror involved here, in these celestial orbitings. In any case, it should not surprise us that the chariot that carries Parmenides to the underworld, as well as the huge doors that open to let him in, should both—at least by the linguistic implications of the word *syrinx*—emit that shrill, whistling, cosmic shriek as the philosopher poet crosses from one world to the other.

If we look back to the Proem, we see rotary motion not only in the chariot wheels and the doors to the Mansions of Night, but also in the more complete rotations of the keys to the gigantic doors held by Justice, as they turn in one direction to open and the other to lock. We also see this motion in a certain image that, though not mentioned in the text, lingers in the mind's eye whenever Justice is spoken of: that of her scales, pivoting around a central balance point. There may even be a quarter rotation when the maidens are pushing back the veils from their faces.

I wonder if there is not also a sense of awe connected to the idea of rotary motion itself: circling about a static core, that empty center containing *non-being* and stillness, while the rim that is rotating is all motion and everywhere *becoming*—as one point moves to the next. The only *being* occurs when the whole thing is viewed from outside, or described by the philosopher/poet. Thus, the whistling roar can also come from the friction between the moving and the still, as *becoming* rubs up against *non-being*, invoking the idea that *being is*, which is the primary elemental philosophical finding that Parmenides brings back to us from the underworld.

As a writer, I often feel that friction—between non-being and becoming—as an idea is coming into view. That feeling of being about to have a thought can be so intense that it might as well be shrieking, and often I make proclamations to my beloved not that I have just had a thought, but that one is on the way. Of course, I announce the former as well, with many struts and gloatings. But it is the unholy noise of metal against stone, or metal against metal, of the edge of thought rotating around the empty center, the becoming scraping against the non-being, that makes me call out in alarm.

"The Garden Party"

Keep in mind these ancient visits to the underworld as we turn to a twentieth-century masterpiece, "The Garden Party," by Katherine Mansfield. This is the story of Laura Sheridan, a girl on the cusp of womanhood, who helps prepare for her family's party and then, when it is over, takes the bounty of leftovers to the impoverished and grief-stricken cottagers down the hill. This errand of Laura's is clearly a voyage to the underworld, to talk with the dead. Following the tradition of such visits, she brings gifts, or sacrifices—in this case elegant sweetmeats and sandwiches—and comes away with new and ineffable knowledge.

The story begins, "And after all the weather was ideal." The first eleven pages of this twenty-two page narrative are taken up with the dazzling preparations for the party. The grounds shimmer with new-mown grass and daisies; hundreds of roses have blossomed overnight. The earth appears blessed; its flora, pregnant and holy, "bowed down as though they had been visited by archangels."

Mrs. Sheridan has decided that her children should take over the running of the party this year. When the workmen come to put up the grand tent in the garden, Mr. Sheridan and his son Laurie have already left for the office, Meg has just washed her hair and has it up in a green turban, and Jose is in her usual silk petticoat and kimono jacket. So, it is the youngest girl, Laura, still eating her childish bread-and-butter, who goes out to confer with the workmen. Unsettled as to her status, and unused to talking to laborers who are not her family's servants, her diction shifts and cracks like an adolescent's voice.

In my view, Laura and Laurie are twins—why else would one name them thus? They seem to me to be sixteen or seventeen, and, in the way of upper-class New Zealand of the early 1920s, Laurie is already a man, going off each day into the real world of the office with his father, while Laura remains sheltered and somewhat infantilized among the women.

The Sheridans are rearranging everything so as to increase delight: the tent is erected, the piano is moved—making a chuckling noise—the whole world seems golden and alive with "soft, quick steps and running voices," with faint playful winds, and with warm reflections of the sunlight from all the silver. Into all this plenty, more is brought: two trays of pink canna lilies, "wide open, radiant, almost frighten-ingly alive on bright crimson stems." At the piano Laura's older sister Jose practices her song—"This Life is Weary"—singing it with dra-matic mournfulness and gusto, enjoying it despite, or because of, her complete lack of experience of the melancholia it portrays.

Almost deliriously sensuous are these preparations, even down to writing out the labels for the fifteen kinds of sandwiches—cream cheese and lemon curd; egg and olive—and then the tasting of the cream puffs from Godber's pastry shop.

Jose and Laura are just licking the whipped cream of those puffs from their fingers when death appears at the back door. Halfway through the story, at the bottom of the eleventh page, the delivery man from the pastry shop tells the Sheridans' cook, the maid, and the manservant that an accident has happened. A man who lives nearby has been fatally thrown from his cart when his horse shied at a steam engine.

Laura is aghast: "But we can't possibly have a garden party with a man dead just outside the front gate."

Though not geographically correct—the man lives in the shanty town down the hill—this is emotionally correct. It precipitates Laura's sudden awareness of death looming. The next five pages of the story are given to arguing about the party and death. Laura's sister Jose accuses her of undue sentiment, and their mother chides her to use common sense rather than emotion, arguing that if a normal death had occurred among the poor cottage dwellers, and the Sheridans had not happened to hear about it, they would still be having their party. Finally, Mrs. Sheridan takes off the large new hat she has been trying on and bestows it on Laura in a sort of rite of initiation to adulthood:

> the hat is yours. It's made for you. It's much too young for me. I've never seen you look such a picture. Look at yourself.

But Laura refuses to look in the mirror. Mrs. Sheridan accuses her of being extravagant and a spoil sport: "People like that don't expect sacrifices from us." Uncomprehending, Laura walks into her own room and catches herself in the mirror, unawares:

> this charming girl in the mirror, in her black hat trimmed with gold daisies, and a long black velvet ribbon. Never had she imagined she could look like that.

Against Laura's will and her beliefs, her mother's surprising gift of the hat pushes aside the thought of death, and Laura decides to "remember it again after the party is over."

Laura does make one more attempt to get advice, from Laurie who has come home from the office, but he exclaims so enthusiastically about her hat that she ends up not asking him what to do about the death.

The actual party in the garden is over before we know it, almost before it begins, taking up less than one page out of the twenty-two pages. The guests,

> like bright birds that had alighted in the Sheridans' garden for this one afternoon, on their way to—where? Ah, what happiness it is to be with people who all are happy, to press hands, press cheeks, smile into eyes.

One of these guests tells Laura she looks "quite Spanish" in her hat, which I take to mean romantic, sexual, foreign—Spain being exactly antipodal to New Zealand. Laura basks in the compliments and offers tea and passion fruit ices. "And the perfect afternoon slowly ripened, slowly faded, slowly its petals closed."

❧ ❧

The afternoon is a flower, and the party itself only a small stopover on the migration of the birds, just as human life, of the individual or the species, is a moment in the life of the cosmos. The adverb "slowly" repeats as a refrain until it seems less like a description and more like a plea, on the part of the narrator or the author. The party takes up such a small portion of the whole story that we begin to wonder if life itself might be what is referred to in the title: Life is the garden party we prepare for, experience, reel away from. We grow into our adult costumes, revel in them, and just down the hill is death.

After the guests have gone home, the Sheridans sit down in the deserted tent in the garden. Mr. Sheridan munches on leftover sandwiches and brings up the news of the death of the carter. Mrs. Sheridan now has "one of her brilliant ideas." She takes the leftover sandwiches, cakes, cream puffs and heaps them into a big basket and tells Laura to take it down to the grieving family of the dead man. Mrs. Sheridan considers adding lilies, but their sexual parts would stain Laura's dress, so no lilies. She starts to give motherly warnings "don't on any account—." She trails off and does not finish, not wanting to plant ideas by actually mentioning the pathways of contamination she is fearing: ingestion, dirt, and sex.

Above all, though, this maternal unfinished sentence warns against contact with the numinous. The reader knows, as Mrs. Sheridan may not, that Laura and her brother Laurie have already explored together this "disgusting and sordid" terrain, because "one must go everywhere; one must see everything." But they had ventured there together, and now, for Laura, it will be different on her own. Laura, who we first encountered gobbling her childish bread-and-butter snack in front of the workmen, now—with her lace frock, her exotic hat, and her basket full of bounty—sets off alone like the young goddess Persephone, to the underworld.

> It was just growing dusky as Laura shut their garden gates. A big dog ran by like a shadow. The road gleamed white, and down below in the hollow, the little cottages were in deep shade. How quiet it seemed after the afternoon.

Just as in classical representations of the other world, the distinctions between men and women, men and animals, are lost here, too, when Laura crosses "the broad road" to the dark and smoky lane, the women are wearing men's caps and the shadows of people in the cottages are "crab-like." The people seem deformed: the old, old woman with a crutch puts her feet on a newspaper as though reading matter were for the feet, not the head. Laura is nervous, wishes she were dressed anyhow but the way she is, wishes to be away, utters a prayer "Help me, God."

As is often the case in the other world, language, too, is distorted. The widow's sister does not understand or respond to what Laura wants or says, and instead of letting her turn back at the Scotts' doorstep, the woman acts as a psychopomp, a conductor of souls, leading her further and further into the house of death. In the heart of the house, the smoky wretched kitchen, Laura finds Mrs. Scott, the new widow, beside the fire. Mrs. Scott is beyond language of any sort. She is a gorgon of grief:

> Her face, puffed up, red, with swollen eyes and swollen lips, looked terrible. She seemed as though she couldn't understand why Laura was there. What did it mean? Why was this stranger standing in the kitchen with a basket? What was it about? And the poor face puckered up again.

In her desperation to flee this Medusa, Laura goes back into the passage, opens a door, and finds herself in the bedroom with the dead man. Again Mrs. Scott's sister mistakes her wishes and draws back the sheet, uncovering Mr. Scott's face.

> So remote, so peaceful . . . he was wonderful, beautiful. While they were laughing and while the band was playing, this marvel had come to the lane. Happy . . . happy . . . All is well, said that sleeping face. This is just as it should be. I am content.

It feels as though the archangel who visited the bushes in the beginning of the story has passed by again. Laura is overcome by the peace and beauty; she feels the need to cry and to say something to the dead man. She sobs, childishly, and then says the wonderfully bizarre and fitting prayer, "Forgive my hat." This is properly reminiscent of "forgive us our trespasses," for her hat was the vehicle of Laura's forgetting about the death for a while. It is also the extravagant insignia of her own coming into sexual flower, a mark of her liveliness and fertility, as well as her ability to travel back to the world of the living.

Having looked death in the face, having sobbed and spoken to it, Laura can now find her way out of the house without the widow's sister to guide her; at the boundary of the underworld, she finds her brother Laurie waiting in the shadows. He had not accompanied her, knowing that she had to make this descent alone. Laurie, being a male, being exposed to the world every day, must have already made the solo journey, but now that Laura has done it too, he can escort her home. He asks her if it was awful, and she replies that it was marvelous, but her experience and her new insight are so far beyond her capacity for words that all she can do is stammer, "Isn't life . . . isn't life?"

Laurie, her twin, having gone there before, understands completely, and is only able, himself, to say, "Isn't it, darling?"

I noted above that the language of the other world differs from the language of our world in that often it is distorted, unintelligible, gibbering, or sibylline. Often it needs an interpreter. Laura and Laurie, standing on the boundary of that other world, are not able to be completely coherent, and the adjective that she needs, that they both need, is not utterable. For any single word would be too finite, too constricting, and too mortal for the concept that she is struggling to pronounce. As though in answer to her mother's unfinished sentence warning her to avoid all contact with the numinous, Laura's echoed broken sentence to her brother is her way of demonstrating that she has, in fact, just collided with it.

Here is another way of looking at their enigmatic words: Having gazed on death in the underworld, Laura reverses Death's negating statement, "Life Is Not," to give the affirming and infinitely open-ended question, "Isn't Life?" This is perhaps her version of what Parmenides finds in the underworld, his brilliant and puzzling discovery that "Being Is."

At the end of the story Laura and Laurie stand in each other's arms at the junction of the lane and the broad white road that separates the world above from the domain below. In the darkness her black hat and its black velvet ribbon would probably no longer be visible, but the daisies circling its rim would glow in whatever light remains, seeming to form a crown of gold.

The Other World of Fiction

It is time for me to revisit my claim that our entrance into the other world when we read fiction is in many ways analogous to the hero's descent to the underworld, or crossing over to the Other World. I base this on the three qualities that seem most indicative to me of such journeys: the disappearance of boundaries, the distortion of time, and the distortion of language.

If the fiction is good enough, we lose all sense of what is going on around us, how much time is passing, who is talking in the next room. We become and stay entranced, losing our personal sense of time as narrative time takes over.

Like dream time, narrative time is non-linear, looping when it wants, disappearing when it chooses. It is elastic, stretching and contracting, two minutes can take several pages, while one sentence may leap thirty years. Narrative time can go in any direction, with flash-backs and flash-for-wards imposed at will. In this way, narrative time echoes mind time, or the time of our imagination, where—although our bodies are embedded in actual time—our thoughts flicker, multi-layered and multi-stranded, looping back, not at all one at a time, but singing to and over and through one another like birds in the forest at dawn.

Tom Lux, in his poem "The Voice You Hear When You Read Silently," says that this voice is "your voice/caught in the dark cathedral/of your skull," and that the way you hear that voice is the product of all experience and feeling, that it is "the clearest voice: you speak it/ speaking to you."[20]

But for me language can feel distorted when we read, partly because we are taking it in through our eyes instead of through our ears, and partly because the author's and/or narrator's voice is so strongly married to our own that we no longer hear with our own mind's ear. The voice we hear when we read silently does not sound like the voice we hear when, say, we think. Certainly our silent reading voice is, as Lux says, based on who we are and what we have experienced and felt. But I think it is some sort of new mixed voice, not solely our own, and this coupled voice becomes the means for inhabiting someone else's thoughts.

Thus, when we are reading, a crucial boundary becomes indistinct: the border of the self has disappeared.

When reading dialogue, we hear voices that are not there. And we view the world described in language that may be different from our own: more brilliant, more Russian, more crazed, more adventurous, more convoluted—and we are made privy to perceptions that are often more vast than our own limitations permit. But to what extent these perceptions will be revealed to us depends in part on what we bring. The hero is always asked the Traveler's Questions, and often asked for other news as well. He is asked to inform his questioners, to describe his provenance, his journey, his motivations. So, too, the experience, thoughtfulness, and intellect of the reader acts, in analogous fashion, to *inform* the text, to impart intelligence, if you will, to the text, as well as the other way around.

When we read fiction, we are under authorial instruction; the author both carries us into and guides us through this new territory. That the writer has survived to tell us the tale allows us to assume that we, too, will be able to make the journey and return safely to ourselves. Thus, under authorial instruction we can travel to places and minds where it might not feel safe to venture on our own. The capable author has made a construction with solid walls; sturdy floorboards cover the cellar hole, and it is now safe to walk there.[21]

Though the reader luxuriates in this safety of following the designated narrative path, the writer often has a more anxiety-filled trajectory. One of the terrors of writing fiction is that at times it feels as though we are constructing those floors with only the flimsiest knowledge of how the foundations have been or should be shaped and poured. As we step out over the void, planks seem to give and sway;

we are not sure if they will hold our weight. A certain queasiness can set in.

In fiction as in heroic katabasis, there is constant passage from the land of the living to the other world of the imagination. The rubbing out and subsequent redemarcation of boundaries can bring a pleasurable tension. It is as though the stack of long lines of the fictional text act as an open door inviting the reader to enter and partake, or at least to eavesdrop. With the traditional novel this form of the text on the page is or has become transparent, inviting us into its trance.

Poetry can also do this, but at times the form—the more architectural slit made by the narrower stack of poetic lines on the page—can feel more like a window, of stained glass or clear, allowing our gaze but not our bodies, keeping us always conscious of who is the architect, who is the viewer. In this latter case, the reader may stay much more conscious of the self: this is the poet and/or speaker over there, and this is me over here, and this is how my mind and imagination react with his, or with what he says. It is more like an ongoing conversation, and the more brilliant or effective the poet, the more I am aware of that other mind, that other soul. With some forms, too, the stanzas (from the Italian *stanza*, for room) keep us shifting among windows into different rooms. The boundary conditions remain crucial and conscious during the reading, and the ensuing trance becomes a sort of *lucid* dream.

In my view, the more extreme forms of post-modern fiction purposefully keep us readers outside, or let us in and then repeatedly, sadistically, expel us. This has little to do with the flow of conversation between equals that takes place between the reader and the writer in lyric poetry; instead, the post-modern fiction writer is always pointing to himself, saying: Look at me! Forget about the damn characters! Here is what I think of the nature of *story*! Here I am! Here is where I am placing you, whether you like it or not!

≈ ≪

We want and demand to be taken in, completely. But we also want, at the end, to be able to take ourselves safely out of the confabulated world, to put it from us. Though it may *haunt* us, it does not drown us, and we can separate ourselves from it, sometimes with effort. That is, as fiction readers we are only comfortable if, when we close our book and gather our wits about us, we can recognize the boundary between the worlds. (Certain mental conditions leave one unable to read fiction because this boundary disappears and the reader is unable to get back to the normal world with any certainty.) Generally we want to be affected, moved to howls of laughter or passionate weeping—this can be a sign of great work—but not permanently devastated.

In the writing of fiction, as well, this boundary between our text and our world can easily become, for longer or shorter periods, uncomfortably permeable. Because the close, focused attention of the writer is often even more piercing and prolonged than that of the reader, the imagined world replaces by its intensity and brilliance the ordinary world of the living. This, too, can lead to a certain terror.

Thus we descend into the book, tumbled into that underworld by the unstable geometry of openings in general and by the particular instabilities invoked by any given opening. In the ensuing silence and stillness of reading to ourselves, under the guidance of a master-seer, we enter a state in which our sense of time may get distorted and language takes on new properties as we hear ourselves think in a voice not our own. This incubation in the cavern of the text echoes the hero's trip to Hades, and if we are properly prepared, if Right and Justice are in conjunction, deities and shades and prophets may decide to show us visions and tell us of, or reveal to us, the mysteries.

III | Forbidden Looking

The Story of Orpheus and Its Condition

It is generally forbidden for living mortals to see or visit the land of the dead; this stringent and fatal prohibition is disobeyed by only a few ancient heroes, many of whom have divine lineage. Odysseus, for example, is the great-great-grandson of Zeus; Heracles the half-mortal son of Zeus; Christ is the son of God; and even Gilgamesh is two thirds god and one third human. Orpheus, who makes the descent and seems to both get and lose what he came for, is the son of Apollo and Calliope, the Muse of epic poetry.

The most elaborate versions of the myth of Orpheus and Eurydice come to us from the Roman poets Virgil and Ovid, and like the hundreds of more modern versions, their accounts overlap and diverge in ways that illuminate the story while reflecting the peculiarities and obsessions of each author. In this chapter, I propose to look at the nature of looking, *forbidden* looking in particular, with reference to the story of Orpheus and how he is seen by Ovid, by Virgil, and by the Flemish baroque painter Peter Paul Rubens (1577–1640). By triangulating from these three views, a fourth, my own, will become apparent.

Both Virgil and Ovid agree that Orpheus, from Thrace in northern Greece, is supreme among poets and musicians and is an enchanting singer. Shortly after his wedding to the beautiful Eurydice, his bride is killed by a venomous snake who bites her on the foot. Orpheus grieves until he cannot stand it anymore, at which point he descends to the Land of the Dead to ask the gods Hades and Persephone to give Eurydice back to him. The song of Orpheus is so beautiful and moving that the whole kingdom of Hades is spellbound: everything comes to a stop, all punishments and pursuits momentarily suspended. Virgil says that the three mouths of the Hell-dog Cerberus forget to bark; according to Ovid, even the Furies weep.

In *Metamorphoses*, published in 8 AD, Ovid gives us Orpheus's passionate argument in full. All the music of his love and grief is in the text. Orpheus recognizes that all humans eventually come to live forever in the realm of Hades, but asks for the favor of letting his Eurydice return to the land of the living for just a mortal while. As soon as Hades and Persephone agree, Eurydice is summoned. She arrives still limping from the fatal wound.

> The Thracian bard accepted her, together
> with the condition set for her release:
> that he may not look back at all, until
> he'd exited the valley of Avernus,
> on pain of revocation of this gift.[1]

We know this story. We know that Orpheus will lead Eurydice out into the darkness of the steep path back, which can perhaps be thought of as a long slanting threshold to the world above. Ovid puts the opening at Avernus, the lake north of Naples, where one of the Roman entrances to the Land of the Dead was located. We know that despite

Figure 11. *Orpheus and Eurydice with Hades and Persephone*
by Peter Paul Rubens (see Plate III).

the "condition set for her release," Orpheus will look back, and thus tragically lose all that he went there for.

But is it at Eurydice that Orpheus is not supposed to look? Or is it at the realm of the dead? Virgil implies that he is not to look back *at Eurydice*, and this is how we remember the story:

> Eurydice, whom he was bringing back,
> Unseen behind his back was following—
> For this is what Proserpina had commanded—[2]

Ovid, however, may be implying that Orpheus may not look back *at the whole realm*, with the phrase: "he may not look back at all."

Whichever the terms of the agreement, we know that our heroic poet will look back and thus lose his Eurydice. Oh, Orpheus, we think, you silly boy, sick with love-greed. Was it so difficult to obey? The Greek gods tended to keep their bargains, why couldn't you trust them and keep your gift? Was it so hard to believe that Eurydice would make it up the precipitous slope right behind you to the world of the living?[3]

How Rubens Sees Orpheus

There is a painting by Peter Paul Rubens in the Prado Museum that shows Orpheus and Eurydice just as they are leaving the throne room of Hades and Persephone (Figure 11). I find this painting as strange and as forceful as Rilke's poem "Orpheus. Eurydice. Hermes." I won't discuss that poem here, except to point out that like Rilke, Rubens gives us a startling view of the Orpheus story, at once disturbing and revelatory.

Looking at the Painting without Knowing

Here is an exercise: Pretend that you do not know what this painting is about. Look at it with eyes fresh and innocent and unknowing; ask yourself what is going on.

We see the older seated couple, and the younger couple on the run, but the body language would suggest a totally different narrative from the one we expect. The girl is still in a state of hurried undress, and her boyfriend, grim faced, is yanking her, perhaps against her wishes, towards the edge of picture. Could it be that the young girl, Eurydice, has been up to some sort of misbehavior with the divine underground couple? She looks back at them with such longing. Has there just been some sort of erotic threesome?

It is easy to read questions into Eurydice's mind: "Is this what has to happen now? Must I really go? Must I leave you?" And she has other questions, private questions for Persephone, who raises her hand in an enigmatic cautionary gesture, perhaps telling her to depart quickly, perhaps asking her to linger.

Eurydice seems almost like the mirror image of Persephone: the young girl being pulled away from the underworld, looking back questioningly at the queen who was once dragged down here, abducted from the world above. Though, of course, neither tale is that simple. I do not think the story of Persephone is one of simple rape. Not only does she stay (for part of each year) with her abductor, but also she is seen and generally portrayed as powerful, knowing, and deeply erotic. At times I wonder if in some way she had abducted Hades and made

Figure 12. Persephone and Hades, Attic red-figured kylix.

him carry her off. In any case, it would be wrong to read her as merely a victim for she is also a goddess and queen. We see this in the red-figured kylix in the British Museum (Figure 12), where Hades semi-reclines on a pillowed daybed, his signature horn of plenty by his side. He extends towards Persephone a bowl of the same form as the one we are looking at, while she holds something in her fingers: has she taken a fig or sweetmeat from the bowl? Is she about to feed it to him? They gaze at each other, each in a state of offering and receiving, erotic and sacramental.

Back to the Rubens Painting: Composition at First Glance

Two strong diagonals make a V shape that informs the Rubens painting. One diagonal is made by the angle of the fleeing bodies of Orpheus and Eurydice. The other is the dark line of the limbs and staff of Hades. It is as though the sight lines of classical perspective have been tilted and instead of leading to the horizon they now lead straight down to a new vanishing point in the abyss. This point is somewhere below the hovering left foot of Hades, below where his staff and Eurydice's left foot meet, and it is appropriately hidden from our view.

Light and Dark

On the murky right side of the painting, the King and Queen of the Dead sit on their three-stepped marble throne, their three-headed dog, Cerberus, beside them. The royal couple looks fulsome, foursquare, rooted, and dark. Fires in the background behind Hades accentuate the shadowy obscurity of their throne. Rushing away from the darkness, Orpheus and Eurydice lean toward the bright world of the living. *We're out of here,* say their bodies. As she hurries, Eurydice's long hair flows out behind her; so does her gauzy dress, which is white like the ghost she is and will remain.

The brightest part of the painting is the luminous body of Eurydice, but breasts, limbs, and feet are highlighted as well. The filmy layers of Eurydice's dress peel from her torso like petals, showing that creamy opulent flesh that Rubens is so fond of. The crimson of Orpheus's tunic, which has slipped down to his waist, is the only primary color; this vivid red signals his status as the only living mortal. His stride is wide, while Eurydice steps tentatively. Her foot hurts. Also, she is turning back for one more look at her hosts.

Hands and crotches occur in a narrow band across the middle of the painting, the breasts in a band above. Breasts are everywhere. The men's nipples are erect. Framed or shrouded by her black dress and veil, Persephone seems sly, wise, and sexy, refreshed and full of appetites. Overbosoming her black bodice, Persephone's breasts are very ripe, catching the light like melons, echoing the fertility of Eurydice's light-struck flesh.

What Rubens Knows about Looking

The scene that Rubens has captured is pivotal, trembling with possibilities. This should be a moment of sudden and greatest hope. Death is on its way to being reversed; the outer world lies ahead, if only one can move quickly enough so that Hades and Persephone do not change their minds. The Orpheus story is a story of love and longing, of poetry and death and Eros. (Many of the Greek vase paintings of Orpheus have the winged figure of Eros flying about the shoulders of the poet.) But above all, this is a story of forbidden looking. Rubens knows all about ways of looking, and it is informative to follow the gaze of his four characters here.

At first glance it seems simple: Orpheus faces where he is going, towards the exit. Eurydice looks at Hades and Persephone. Hades looks at Persephone. Persephone looks off into space. Thus all the gazes are skewed. The atmosphere is full of smoke and partly hidden fires.

The gaze structure is more complex than this, however, for as Orpheus faces the outer world, striding out of Hell and out of the picture frame, he rolls his eyes back to (try to) keep watch on Eurydice. He is already looking back, already disobeying, and he has not even left the throne room of Hades! Eurydice looks at Persephone and Hades, full of questions and longing. Persephone peers off not just into space but into time, into the future like a seer, knowing and sad. Hades, sitting not quite squarely but contorted, left knee over right, left arm over right, holding his staff against his knee, looks back at his wife, questioning and alarmed. This is a fleeting moment, made more intense by the fact that the pose adopted by Hades is hard to hold for more than a few seconds. Try it. This odd contortion of Hades echoes the more tightly twisted pose of the seated Christ, called "Man of Sorrows," as seen on the title page of *The Small Passion* of Albrecht Dürer, dated

1511. Auguste Rodin will later use a form of this pose for his *Thinker*, in 1880, as part of his portal of *The Gates of Hell*.

The gazes of our quartet form a cat's cradle of intersecting and crossed regards. The hands are all erotically charged. Eurydice, in such a hurry that her clothes are falling off, frames a breast with one arm. With her other hand she holds her filmy outer garment up to her crotch, and Orpheus grips her there also, as though by pulling her by the groin he can bring her back to the world of the living. Persephone's right hand hovers in front of Hades' lap, pointing to Eurydice's genitals, while Hades, holding his craggy body in that untenable position, holds his long spear with thumb and two fingers in a suggestive manner. Eurydice, naked down to her haunches, looks back at Persephone and Hades. It is clear now what her questions are: Will I ever know such erotic splendor? Will I ever know such divine thoughtfulness?

How Virgil Looks at Orpheus

Virgil's *Georgics*, written between 37 and 30 BC, make up an ecstatic farming manual. The four long poems of the *Georgics* are in the form of instructions having to do with the tilling of fields; the tending of vines and fruit trees; the raising of livestock; and the keeping of bees.

In the fourth Georgic, Aristaeus, son of Apollo and the water nymph Cyréné, is in despair because his stock of bees is ruined. Cyréné tells him to go ask Proteus, the prophetic shape-shifting sea god who must be captured and held onto as he changes form. Proteus tells Aristaeus that his bees have died because Orpheus is angry at him for chasing after Eurydice with amorous intent just after her wedding— for it was in trying to escape his advances that she stepped on the snake that fatally bit her. Proteus goes on to recount the grief of Orpheus, his descent to Hades where the flittering shades congregate to hear him sing. Although Virgil leaves out the entire content of Orpheus's song, the pleading and the arguments, he gives a marvelous description of its effect on the inhabitants of the underworld:

> Spellbound the snakes in the hair of the Furies too;
> And Cerberus the Hell-Dog's all three mouths
> were open-mouthed and silent, forgetting to bark;[4]

Then Virgil describes the ascent of the young couple from Hades: everything is going carefully and well when "a sudden madness" grips Orpheus, and

> . . . seized by love,
> Bewildered into heedlessness, alas!
> His purpose overcome, he turned, and looked
> Back at Eurydice! And then and there
> His labor was spilled and flowed away like water.

Eurydice calls out to him, at first angry,

> "What was it," she cried, "what madness, Orpheus, was it,
> That has destroyed us, you and me, oh look!"

And then she whimpers to him in pitiful and guilt-provoking terror,

> The cruel Fates already call me back,
> And sleep is covering over my swimming eyes.
> Farewell; I'm being carried off into
> The vast surrounding dark and reaching out
> My strengthless hands to you forever more
> Alas not yours."

Orpheus is left, "Clutching at shadows, with so much still to say."

Virgil's tragic and single-stranded story is the one we think we know. With his telling ringing in our minds, we decide to restrain our greed. We vow to embrace patience and reason and obedience, to settle for small and manageable amounts of love, in order not to lose what we have. We decide to watch our footing in the present, to look to the future, to trust in trust.

But we should remember who Virgil is, and what his obsessions are. His real passion here is for the work of husbandry and the labor of farming, and these four poems, while they can be ecstatic and celebratory, are also full of gloom about natural and political disasters. Although at times they read like a *Farmer's Almanac* of the sublime, the *Georgics* remain on some fundamental level a manual of instruction. Virgil recognizes Orpheus as an enchanting singer, but underlying his whole tale is his concern for notions of work and doing things correctly. Thus, he is highly critical of Orpheus's work habits—"*His purpose overcome*, he turned, and looked / Back at Eurydice!" (emphasis added)—and he describes the grievous result of losing one's beloved in terms of farm work: "his labor was spilled, and flowed away like water."

But never has there been someone more unlikely to follow instructions than Orpheus. He is a genius, a poet, a musician, not a farmer, and his instruments are the imagination, language, and the lyre, never the plough. Descended from and inspired by the Muses, he is not one for prudent behavior or stolid obedience.

The Difference between Looking and Seeing

But what is *looking*, this action that is so forbidden in the Orpheus story? How can something that we do with our eyes be so crucial and so fatal to the one we gaze upon? What is *seeing*, as opposed to looking? What about seeing as understanding?

There is a *felt* experience of looking that is different from that of seeing. When we look, it really feels as though we are sending rays from our eyes to the object of our inspection. We know when we are looking. We feel it in the muscles that turn and that adjust the focus of the eye. We feel it inside as well. Seeing, however, is what our eyes tend to do naturally, passively, without feeling.

So too with listening and hearing, though the sensory difference is subtler. Still, when we listen for a sound just at the edge of the audible, it feels as though something is going on inside the ear that if we were a cat or an elephant would probably have a clear and visible manifestation: a swiveling, a pricking up, a change in curvature of the flap—as though we could squint and focus with our ears.

Looking is intransitive. Thus we must look *at, for, under, inside, around*, and so on. Such prepositions distance us from the object of our attention. It is as though we need to know where we are, by means of these prepositions, because we are not yet achieving, but are still in a state of wanting, full of purpose and intention, to see.

Seeing, however, is transitive: when we see, our object falls into our gaze, sometimes as the result of looking and the impulse of our attention, but it can also just *happen*: Something chances into our field of vision, unlooked for and unattended, and we see it. Of course with any luck the seen object takes up residence in our mind and our understanding. Oh, I see. I grasp what you are saying. Seeing with the eye of the intellect becomes almost interchangeable with the ocular version.

This, of course, is one reason why looking can be forbidden: because of the danger of sight leading to insight. Mortals are not supposed to enter the realm of the dead before they die; in doing so they would not only see death, but understand it.[5]

We will come back to the difference between looking and seeing in a moment, but first what about the Evil Eye? What comes out of our eyes when we look?

Eye Beams

The question of whether something comes out of the eye when we see is a very old one. The Greek pre-Socratic philosopher Empedocles (c. 495–432 BC) says that the eye works rather like the lantern carried by a traveler on a stormy night in winter. Such a lantern has linen sides, to keep the wind from extinguishing the fire within. This fire has beams that are fine enough to pierce the linen and shine outside. So too the eye (fashioned, Empedocles says, by Aphrodite) has membranes that keep its fluids inside, but allow the eye's fine fires to pass through to the outside. These fires passing from the eye are the eye beams.[6]

Aristotle has an excellent objection to the notion that we see by means of fires coming from the eye, saying that if we did, then we could see in perfect darkness.[7]

Plato, in the *Timaeus*, has a view that actually counters this objection, for he says that the eyes both project and receive.[8] Thus, according to Plato, eye beams, consisting of smooth fire that shines but does not burn, project out from the eye to interact with "the outer fire of day," becoming "the entire stream of vision" that is "aligned with the direction of the eyes." This stream can transmit certain "motions" back into "the entire body until they reach the soul," thus producing our sensation of vision.

Both the inner and outer fires are necessary: light from the external world cannot penetrate the eye unless it interacts with the fire from within. And if the outgoing beam encounters only darkness, then the eye's own fire is "utterly quenched," the eyelids close sleepily, and the trapped inner fire gives rise to the phantasms of dreams.[9]

The modern physics of optics treats the lens of the eye like any other lens, and the ray diagrams of optics show all the light coming from outside of the body. This light is then focused by the lens to form an image on the cells of the retina lining the back of the eye.

Modern neurobiology, however, has come to realize that vision is not as passive as the ray diagrams imply, but involves much interaction with higher brain centers.[10]

That there should be some active component to looking makes sense, actually, when we consider the felt experience of our everyday life. When we really look, we can feel all of our focused attention beaming out in our gaze. All the intransitive verbs of visual yearning—look, peer, glance, squint, spy, gaze—can have this intensely felt component, and the fullest sort of looking at something, either ocularly or metaphorically, comes when we pour ourselves and everything we know into our observations.

So, seeing can feel very different from looking; seeing can feel simply like lying back on the grass and watching the clouds move across the sky; but seeing can also feel like gathering things in, when after a while we can almost sense the image impinging on our eye. Or, when, after much looking, we finally sense the quiet spread of insight and understanding.

᷒ ᷒

The fullest sort of listening can be analogous to looking. In my Hindustani singing lessons, it feels as though the best listening involves projecting a sort of song from the ears, which joins with the song we are singing and comes back to us.

Our minds are constantly telling our senses what to notice, what is curious, mysterious, astonishing, worthy of our attention. Thus alerted, our senses become like the antennae of the male silk moth—those feathery brown appendages, like ferns, almost as long as the rest of the animal, superbly ready, quivering, able to sense a single molecule over unheard-of distances.

Sometimes it helps to consciously sensitize the self in various ways, to get rid of obstructions to feeling, even to the point of a certain rawness. When I was a silkworm biologist, some of our experiments depended on intricate dissections, where we would cut out different regions of the eggshell of the silk moth. These eggshells are small—about the size of a poppy seed, or, in some species, a sesame seed—so we had to construct our own tiny surgical knives. The old Gillette double-edged "Blue Blades" are the best, being brittle enough to snap into shards if you worry them correctly with pliers. Working under dissecting microscopes, we held the razor slivers with the finest possible forceps. By comparision, our hands—our mighty fingers—were so large and clumsy, particularly when magnified by the scope. If you sandpaper your fingertips a bit before any exploit that needs such exquisite control, however, it is as though you magnify your sense of touch as well.

Touch differs from vision, of course, because of contact. It does not take place over a distance, so it is not as though we need to emit beams from our fingertips. And yet, by sanding off calluses and layers of dead skin, it feels as though we are casting our haptic sensitivities right to the skin's surface, just as kindling the fires of the eyes is part of our necessary preparations for observing the natural world or the world of the human psyche.

Of course, the next step is to do this without the sandpaper. I have seen elephants behave as if their whole body were an exquisitely sensitive hand. Why could the whole body not also work as an antenna for those senses that act over greater distances? Or perhaps, like the male silk moth, we should learn to grow huge ferns from our heads, reaching down to our knees except when alerted, wings of the senses.

The scientist would say, Eye beams or ear beams, what nonsense: it's all a matter of focus and concentration; it's all in the muscles in and around the eye, or the ear, as well as in the higher brain centers.

But that is not what it *feels* like. You know what it feels like. It feels like throwing the fire of your self out through your eyes. Whether you look at your beloved or at a tree in bloom or at a rare wild poppy you have just stolen.

❧ ❦

Consider eye contact. We know when it is happening. It is a very precise state. We even know when we are making eye contact with a cat. Is there a set of subliminal markers that let us know when eye contact is happening? Do the pupils simply dilate or contract? Why does it sometimes feel like we are falling into the other person?

Face to face with a statue, or a dead person, we don't feel this.

The Evil Eye

Part of the danger of looking and being looked at has to do with the Evil Eye. This notion depends, of course, on the conception of the eye as emitting a vicious and life-sapping force, rather than a gentle fire, or the infinite and enlivening mirror-gaze of the beloved.

On my desk as I write is a newspaper article dated February 28, 2008, with the headline "Italy: Men Can't Grope . . . Themselves."[11] It begins: "Whatever their reason might be, a passing hearse or simple discomfort, Italy's highest court ruled that men may not touch their genitals in public." While the defendant's lawyer argued there had been a problem with his overalls, "the court struck against a broader practice: a tradition among some Italian men of warding off bad luck by grabbing the crotch." This warding off of bad luck is a classic instance of protecting against the Evil Eye.

In his article, "Wet and Dry, the Evil Eye," the anthropologist Alan Dundes says that when the Evil Eye is at work, buildings or rocks fall apart, people or animals fall ill with loss of appetite, excessive yawning, hiccoughs, vomiting, fever, or impotence. If the victim is a cow, its milk may dry up. Plants and fruit trees suddenly wither and die.[12]

The Evil Eye is, Dundes points out, connected to envy, even in the earliest Near Eastern texts; the English word *envy* comes from the Latin, *invidia*, from *in videre*, thus from the verb "to see." To explain the seemingly disparate details of the Evil Eye complex, for example, why "males often touched their genitals upon seeing a priest or other individual thought to have the evil eye," and why phallic gestures like the *fica* are used for the same purpose, Dundes links the notion of envy with the idea that the crucial life-giving liquids, including body fluids, exist in limited amounts, and the more one person has, the less there is available for others.[13]

Thus, when the blood-thirsty dead envy the living, or when the old envy the young, or the barren envy the fertile, according to Dundes, they are after blood or maternal milk or semen or simply the sap and vitality of youth. Any overt praise or admiring glance may indicate a wish for such precious fluids. "If the looker or declarer receives liquid, then it must be at the expense of the object or person admired." Hence the withering, the drying up.

❧ ❦

In some sense, Orpheus looking back may have been unnecessary. For would not Eurydice have been the victim of the evil eye of *all* the dead who were not given her chance of being similarly revived? Or could Orpheus have been resentful and envious of her lack of grief, for he is the one who feels and sings his pain, both times when she dies? There is an uncomfortable feeling that she gets off scot-free. Even in Rubens' painting, Orpheus is grim faced, while Eurydice seems full of longing, not for him, but for her new chthonic companions.

Getting around the Forbidden

The phrase *forbidden looking* can mean two different things: The first is that we *must* not look: i.e., it is against some law, moral code, practice of physical or psychological safety, convention, or taboo. The second is that we *cannot* look: that is, the laws of physics or metaphysics prevent us from seeing. The laws of physics may keep us from seeing because the object is too distant, too large, too small, too fleeting, or too dark. The laws of metaphysics may keep us from seeing because the object is too infinite, too eternal, too complicated, too devastating, or in some other way incommensurable with human understanding.

As we shall see, there are some objects of examination that are forbidden in both senses, or for which there is an oscillatory shimmer between the two forbiddings. In fact, things forbidden by moral code or taboo may be that way because of their metaphysical status. Danger and impossibility are often linked.

Skewed Looking

There are various ways of getting around some of these prohibitions. For example, some things are impossible, or too dangerous, to confront until we have undertaken suitable preparation. The religious mystic bumping up against the divine comes to mind here.

Sometimes, with the metaphysically dangerous, it is best to do something akin to what Emily Dickinson calls "Telling It Slant":

> Tell all the Truth but tell it slant —
> Success in Circuit lies
> Too bright for our infirm Delight
> The Truth's superb surprise
> As Lightning to the Children eased
> With explanation kind
> The Truth must dazzle gradually
> Or every man be blind — [14]

Plate I. *Christ in Limbo* by Fra Angelico (see page 2).

Plate II. *Pythagoras Emerging from the Underworld* by Salvator Rosa (see page 18).

Plate III. *Orpheus and Eurydice with Hades and Persephone*
by Peter Paul Rubens (see page 54).

Plate IV. "The American Cowslip" by Robert John Thornton (see page 82).

Plate V. *Descent into Limbo* by Dionysius (see page 97).

Plate VI. *The Descent into Limbo* by Duccio di Buoninsegna (see page 98).

Plate VII. *Christ's Descent into Limbo* by Jacopo Bellini (see page 100).

Plate VIII. *Christ's Descent into Limbo* by Martin Schongauer (see page 102).

Plate IX. *The Harrowing of Hell* by Martin Maler (see page 103).

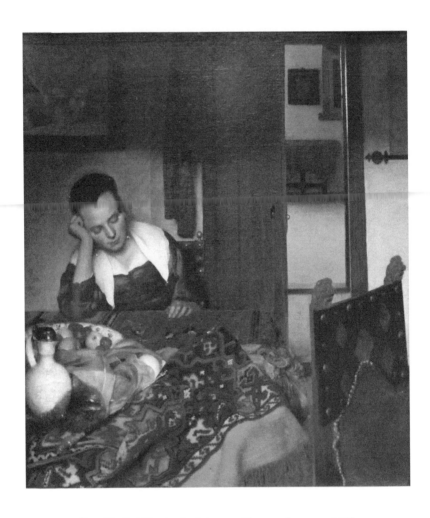

Plate X. *Girl Sleeping* by Johannes Vermeer (see page 129).

When it is physical laws that are preventing us from seeing, we have to use various forms of skewed looking. We cannot look at the sun during an eclipse or it will blast our retinas. So we obscure it with smoked glass; or we hold a baseball cap above a sheet of white paper and use the five grommeted holes as diffraction lenses to focus the image of the progressively moon-obscured sun on the paper;[15] or, as an old Chinese woman I met on the day of a solar eclipse explained to me, "you bundle up your sweater in front of your eyes and look through it into a pail of water in which the sun is reflected." We use radio telescopes to see the farther galaxies; we bounce x-rays off of the denser parts of our internal organs; we fill bubble chambers with superheated liquid hydrogen to look at the behavior of electrically charged subatomic particles; we use magnets to focus beams of electrons to look at the microscopic components of cells; or we look in convex mirrors to help us see when we are backing out of a driveway—all types of skewed looking.

An extreme view would have it that all binocular vision, looking with both eyes, is a form of looking slant, as the brain fuses the glances—one skewed a bit from the left, one a bit from the right—into a centered-seeming view. Of course, metaphor, too, can be seen as a way of using slant images to look at what is right in front of us. As with binocular parallax, it is the disparity between the two images that allows us to perceive depth. Perhaps, too, in the Orpheus story, the disparity between the views of Ovid and Virgil, triangulated with what Rubens shows us, can give us better perception of the depths of what really happened.

Seeing in the Dark

Sometimes skewed looking really means averting our gaze just slightly, for example, when we need to see in the dark.

There is that weird business of looking at stars in the night sky and almost seeing a very dim one, the kind you cannot see by looking straight on, but only with a sideways glance. You have to concentrate on it, but at the same time you have to avert your gaze. The physiological reason for this, coming from the physics of vision, is well known: the visual cells that are most sensitive to dim light—the rods—are more densely packed away from the center of the retina. So, we will be using more of them wherever our gaze is not centered. Though the rods have little sensitivity to color, they are sensitive to motion, so shifting the head slightly can help make the faint object visible.

This is so easy to say and to explain, but what I forget when not actually looking at the night sky is how intensely we are pulled toward looking at things directly. No matter how much I know the science of it, the felt experience remains uncanny. At first I center my gaze just to test if the star is really all that dim. But even when I have proved that it is, I still persist in leaving the slant. However much I know that averted vision is the only way to see a really faint thing, I keep turning to look at it full on. And then it disappears of course, and I disbelieve I ever saw it, until I turn away again. The same pull towards the center happens when I am walking outside on a black night, even though a friend taught me that here, too, I must keep my eyes focused on the brim of my imagined baseball cap, rather than on the uneven ground, in order to see that ground better.[16] There is some ingrained compulsion to bring all of our attention to focus on the center of the mysterious thing. Stargazers, night walkers, mystics, and poets sometimes have to teach themselves to stay their gaze at the edge of itself in order to penetrate that dimness where the interesting things happen.

Perhaps it is forbidden, though no one says this aloud, to watch through all hours of the night. I do not mean watch the stars, but simply watch. It certainly feels as though only the initiate, or the night shift worker, or the night watchman, can look at the night as she disrobes and lies naked under the blackening sky. The reason sleep comes is to keep us from looking at the night in its entirety. And yet, what visions come to one who has watched through all the hours!

Mirrors

Mirrors, too, can be a means of looking at something that is too dangerous to look at straight on. In the Greek myth of Perseus and Medusa, the mirror attenuates what would otherwise be a fatal gazing, for Medusa is an extreme example of the Evil Eye. Her gaze causes no slow withering and desiccation, but death and sudden petrifaction. Her sisters, the other two Gorgons, have similar powers, but they are immortal, while she can be beheaded. The best and possibly earliest full account of Perseus and Medusa comes from the second–century-BC mythologist and historian, Apollodorus.[17]

Perseus is a demi-god, Zeus having come to his mother, Danäe, in the form of a rain of gold coins. Unlike other heroes whose regalia

Figure 13. *Perseus, Athena holding Medusa's Head, Hermes* by the Taporley Painter. (**Photograph © 2010 Museum of Fine Arts, Boston.**)

simply recalls their exploits, Perseus has clothing with magical properties, sandals that enable him to fly, a cap from Hades that renders him invisible.

Perseus flies to the cave where the scaly winged trio of Gorgons are sleeping, and looking only at the reflection in his shield, with Athena guiding his hand, he chops Medusa's head off with a sickle-bladed sword given to him by Hermes.

It feels awkward, in fact, to manage all this: try holding your shield on your left arm, looking into it like a mirror, while at the same time grabbing onto the snakes of Medusa's head with your left hand, in order to steady it so you can cut off the head with your sword that is in your right hand. Perhaps this is why some accounts have Athena guide Perseus' hand.

Medusa's oddness continues after death, for as soon as she is beheaded she gives birth to full grown Chrysaor and his brother, the winged horse Pegasus, both fathered by Poseidon.

The power of Medusa's Evil Eye also continues after her death. On Greek vases, Perseus does not look at her severed head, but only at its reflection in his shield or water. Sometimes Athena holds the head and shield for him as in the Apulian red-figure vase in the Museum of Fine Arts, Boston, where Hermes, holding his staff, looks on as Perseus gazes at the reflected image—which has been reversed, top to bottom, by the optics of the convex mirror of the shield. Perseus wears the cap of invisibility and the winged boots given to him by Hermes. Despite his cap of darkness, we are able to see the young victorious hero. This must be due, I think, to the imaginative genius of the artist. (See Figure 13.)

In Apollodorus' account, Perseus later uses the severed head to turn his enemies to stone, while Ovid shows sea nymphs placing fresh twigs next to the head and having them turn to coral.[18]

The ancient theories of eye beams meeting and mixing and returning to the sender might make one leery of looking at even the *reflection of Medusa's face*, yet for some reason the mirroring shield, or the surface of still water in springs and wells, can weaken Medusa's gaze until it is no longer dangerous.

In fact, Perseus seems to use mirrors much the way we do: to look without looking. With mirrors we can be sly, looking at others without giving them our face, akin to looking through veils. And yet such mirrors, too, are the only way we can see ourselves, unless we trust the reflection we see in others. With mirrors, as with writing, we can see around corners; we bring the lantern where it is not; we shift our vantage point without moving and thus alter the angle and power of perception. This use of reflection can be a form of hinging from absent to present, from *is* to *is not*, from one mind to another. Pythagoras is said to have claimed he could write on the moon by writing in blood on a mirror, and showing it to the moon, whereby his inscription would be seen reflected on the moon's disc.[19]

Two mirrors gazing at each other can have properties that seem to be more than the sum of the parts, giving us an inkling of the appearance of infinity.

Forbidden Mirrors

Sometimes mirrors themselves become forbidden to look into. Frazer, in *The Golden Bough*, describes how this prohibition comes into play when the soul is in danger.[20] In certain cultures, he says, the reflection of a person in a mirror is thought to be the soul. Because of this, sick people should not look in mirrors, because when one is ill "the soul might take flight so easily." It is even worse when someone has just died; then all the mirrors in a house are covered up or turned to face to the wall, for the ghost of the newly dead, presumed to linger about till the burial, may carry off any souls found as reflections of faces in the mirrors.

Of course when the danger is over, these mirrors can again be looked at.

But some mirrors must never be looked at, not ever, at least by humans. This is the case with the sacred regalia mirror, Yata no Kagami, kept at the Ise Shrine in Japan. This is presumed to be an octagonal mirror, ritual in nature, similar to other known octagonal mirrors.

This forbidden mirror is kept in a box within a box within a box, hidden from outsiders in a ritually restricted area of the shrine. The outermost box is made of Japanese cypress wood (Hinoki). The middle box is also of cypress wood. The innermost box, containing the mirror, is made of gold, and thus is incorruptible.

This Shinto sacred mirror is not, in fact, a looking glass, but rather an emblem of imperial nobility in ancient Japan, and, at least in ancient times, possibly a device for reflecting light, thus connected with life and fertility.

No one may look at this mirror, not even the Emperor, though some say the Emperor may have seen it during his pilgrimage to the shrines in 1869; and there is the possibility that Shinto priests may have glimpsed it during a ritual in 1901, when the innermost container was permanently sealed. [21]

꙰ ꙰

What would it mean to have a hidden thing that one must not look at? What if I had such a box in my closet? Would it be like having a wrapped present that I am not supposed to open until a certain day—except the day is infinitely far in the future? How soon would I succumb?

Is it perhaps a position of power, to have such a thing and forbear to open it, except for when one is *in extremis*. I am presuming it would

help, in such a condition. It feels very different from the deferred pleasure of, say, a birthday box, full of surprises, that one is invited to open
at any time between the receiving of the box and one's birthday the
following year. Pleasure finitely deferred is not the same as pleasure
eternally denied.

"What is that?" my friend would say, looking into my closet for
the first time.

"That is the box I never open."

"What is in it?"

"I have no idea; I have never opened it."

"But have you never thought of peeking?"

Of course I have. I have even thought of rigging up a series of mirrors. Or fiber optics. Ways of looking without looking.

It turns out that the sun goddess, Amaterasu Omikami, gave this
Shinto mirror to her grandson, the rice god, when he was sent
down to earth to restore order, saying, "Think of this sacred mirror as
none other than myself, take care of it, and worship it forever." Ever
since, successive emperors have worshiped the mirror and believed the
goddess to reside within it.[22]

Now the business of the box in the closet gets more complicated: it
is not forbidden simply because those are the, possibly frivolous, rules;
there is a deity inside and it can be mortally dangerous, as we shall see
below, to look upon the gods. No wonder the sacred mirror is kept in
a restricted area of the shrine at Ise.

But notice how in both cases, the ancient mirror in its three boxes,
and the hypothetical box in my closet, gain a large part of their meaning and fascination from the taboo.

A friend tells me that when people die on Mount Everest or K2 it
mainly happens after they reached the peak, when they were on
their way down.[23] I do not know if this is statistically true, but if it is,
I wonder if it might come from the obvious fact that when ascending,
the climbers have their goal constantly in view, but afterwards it is

behind them, and thus they look back when they should not, lose their footing, perish. My friend argues that as well as the exhaustion, there is also no more adrenaline, no more point to life, once the peak has been attained. I suggest to her that Orpheus might have been more intently focused on getting into the world of the dead than on returning to the world of the living. She claims, Well of course; how would their life have been: "Shall I shave now, do you think? Where are my slippers?"

Perhaps the highest Himalayan peaks are a kind of trapdoor. Once you have achieved them, life is altered, nothing can be the same. The goal is lost.

Ways of Looking at the Divine

When I was a biologist, I spent my days in darkened rooms, looking at parts of insects through the electron microscope. I was looking at the eggshells of the silkworm; in cross-section their architecture is like some mad postmodern construction of a Greek temple with a castle on top. The planks and beams of this castle are made of protein fibers in a matrix, much like the fibers in fiberglass. Every time I increased the magnification—10,000, 30,000, 50,000 times—I saw more of the strange beauties and hidden order that are revealed only at the microscopic level of the natural world. I was overwhelmed. I would think of the story of a German scientist looking for the first time into one of the earliest electron microscopes, turning to his colleagues, stricken. "Gentlemen," he said. "We are looking up the pant legs of God."

For years I felt his awe, but I was a scientist and I let the strangeness of his image slip right by me, unexamined.

Now that I am a writer, I find the phrase "the pant legs of God" disturbing and wonderful. Look what it says about our point of view and how it evokes an image of us as slithering along the ground, for how else could we look up pant legs? It implies, too, that God has not only attributes, but also private parts, and that we are looking at or towards the forbidden parts of someone who it is forbidden for us to see.

The gods of the Greeks, Egyptians, Sumerians, Persians, and Hindus had bodies, or avatars with bodies; they were highly sexed, if sometimes hermaphroditic, and their stories are full of exploits of engendering. But the Judeo-Christian god appears as pure spirit, sometimes as a voice. This Western god is one of the only deities who does not eat or dance or engage in sex. Sometimes the Hebrew god talks of His body, as we will see in a moment, but interpreters claim even then He is still completely spirit. He seems to use his body more for metaphor than pleasures of the senses.

In the Ancient Greek, Jewish, Christian, and Muslim traditions, the danger of looking directly at deities is well known. We all know that the sun god will blind us. Other viewings can be even more dire, as in the case of Semele, Jupiter's young mistress. Ovid tells how Juno, jealous of her young rival, convinces Semele to test that her lover is really Jupiter by asking to see him, "just as you are when Lady Juno receives you in her embraces and you initiate the pact of Venus." Jupiter reluctantly agrees and goes home to Mount Olympus to pick out a minor thunderbolt, one with "reduced anger and a lower flame." He comes thus to Semele, but her mortal body cannot bear such heavenly excitement, and she bursts into flames. She is pregnant at the time she incinerates, with Jupiter's child, Dionysus. The fetus is taken from her womb and sewn into Jupiter's thigh until it comes to term.[24]

☙ ❧

Despite these perils, even in the Western traditions the religious mystic seeks to converse with and see and have union with the divine—to meet, meet up with, nudge up against, be taken by, be consumed by, become one with, sometimes even to the point of *becoming*, for a moment, the god.[25]

To be taken or taken up by the god is a dangerous matter, physically and psychologically, and preparation for such union is the life work of the mystic. That the imagery used for divine union is often both incendiary and highly erotic is no accident.

Prophets in the Abrahamic traditions often settle for somewhat less than mystic union with the divine. Instead, they speak with and for the deity. Divinely inspired revealers, they often want to know and see god. In the Book of Exodus, Moses, as he leads the Jews out of Egypt, converses with God who appears wrapped in a pillar of cloud. Though they are standing face-to-face, Moses feels that it is really face-to-cloud. He wants more and asks to know God's ways. God answers elliptically, perhaps evasively. But Moses persists, saying, "Show me, pray, Your glory."

It is not that for Moses seeing would be believing, for he believes in any case, and all along he has been talking with God and making covenants with him. Perhaps it is that that *not-seeing* can bring about malaise, as there comes a time in a relationship when the level of intimacy brings with it a feeling that one should be closer, one should be able to "see" more. This seeing is fatal and disastrous with Semele and Jupiter, as we saw above, and it causes great trials for Psyche, when oil from her lamp falls on the sleeping Cupid whom she has been forbidden to look on.[26]

In Exodus it is as though Moses is trying not to be petulant, but he wants and feels that he deserves the full revelation, an exhibition of God's face. God answers in his own language, which is, as usual, somewhat elliptical, saying: "I shall make all My goodness pass in front of you, and I shall invoke the name of the LORD before you. And I shall grant grace to whom I grant grace and have compassion for whom I have compassion." But then He clarifies and extends a divine warning: "You shall not be able to see My face, for no human can see Me and live." [27]

Robert Alter, translator and commentator, says that the implication here is that "God's intrinsic nature is inaccessible, and perhaps also intolerable, to the finite mind of man." This contains both of our meanings of "forbidden": *inaccessible*, due to the physics and metaphysics of incompatibility between man and God; and *intolerable*, because of physical and spiritual danger.

Without letting Moses respond, God then proposes a compromise, saying:

> "Look, there is a place with Me, and you shall take your stance on the crag. And so, when My glory passes over, I shall put you in the cleft of the crag and shield you with my palm until I have passed over. And I shall take away My palm and you will see My back, but My face will not be seen."[28]

In this translation, in the phrase "you will see My back," it is almost as though God is some immense unseeable ship, and Moses can only see its wake, its disturbance in the smoothness. He can know God by inference, from his works and glories. But what we want, the way we

Figure 14. Ceiling of the Sistine Chapel by Michelangelo: Creation of Sun,
Moon, and Planets.

come to know the beloved, is to gaze on the face. This, Moses is not allowed.

In the King James Version, a slight shift in wording of these ambiguous utterances makes for a very different image: "And I will take away mine hand, and thou shalt see my back parts: but my face shall not be seen."[29]

To me, "back parts" has to mean "buttocks," and it begins to sound as though God is mooning Moses. The moon is of course the lesser light, which we may look at with impunity. We might wonder if Michelangelo was thinking of God in Exodus when he paints His naked rump on the Sistine Chapel, right beside the draped frontal view of God creating the sun, moon, and planets (Figure 14).

The *back* of God is such a different thing from the *back parts*; the back leaves us full of yearning, as though we have come on the scene too late and the divine face is just beyond our gaze.

Mohammed

It is more often in legendary works than in canonical scripture that the God of the Peoples of the Book does allow certain prophets to see him. In the *Mirâj Nâmeh*, the mystical legend of the miraculous ascension of Mohammed, the Prophet is taken at night from the sacred mosque at Mecca to the far-off Mosque in Jerusalem, and from there to the throne of God in the seventh heaven.[30] His guide is the winged angel Gabriel, who flies beside him while Mohammed rides on the *Buraq*, a fabulous animal, swift as lightning, with the body of a horse and the head of a woman.

In the Bibliothèque Nationale in Paris, there is a fifteenth-century illuminated manuscript of this amazing journey, with calligraphy in the Uighur script of eastern Turkey. This book is full of stunning paintings that I do not dare show you here, for they depict Mohammed as well as the prophets and angels he encounters on his single night's journey to his audience with God. The Uighurs are a Turkic people of Central Asia who adopted Islam in the tenth century. In these paintings the faces and clothing appear very Asiatic. Mohammed and the other prophets are shown bearded; they all wear long Mongol robes, but Mohammed's is always jade green in color, and he is always enveloped in a prophetic flame of gold. The *Buraq*, his graceful woman-headed horse, has a reddish body with white or gold leopard spots; on her head she wears a gold Mongol crown. Her black hair flows down her long arched horse's neck. The skies are lapis with golden Chinese clouds, coiled and tumultuous. The angel Gabriel has polychromatic wings that spread and gesture in the sky, dazzling and expressive.

So much is shown to us in these paintings, but when Mohammed, who does have, in this legendary account, direct access to God, comes into the Divine presence, all *we* can see is his small turbaned figure, in green robes, bowing down, face to the ground. Except that there is no ground, no sky, only flames: everything is swirling golden flame. God is unshowable, perhaps unknowable, except as golden fire. Through it all, Mohammed keeps his eyes brilliantly, daringly, open. My own copy of this book falls open to this page.

ᕱ ᕲ

Prophets and mystics want to know and see their god, and so do other mortals, particularly those who are intimates of the god in his human form, such as the apostles of Christ. In the apocryphal Gospel of Bartholomew, the apostles question Mary until she finally tells them what it was like when the Holy Ghost announced to her that she would conceive the son of God.

Mary's narration is so mystical that it is right on the edge of forbidden understanding:

> As she was saying this, fire came from her mouth, and the world was
> on the point of being burned up.
> Then came Jesus quickly and said to Mary, "Say no more, or
> today my whole creation will come to an end."
> And the apostles were seized with fear lest God should be angry
> with them.[33]

Jesus takes the apostles and goes to Mount Mauria with them and sits
down in their midst. They are still fearful, but he says, "Ask me what
you wish, so that I can teach you and show you. For there are still
seven days, and then I ascend to my Father and shall no more appear
to you in this form."

Hesitating, they say, "Lord, show us the abyss, as you promised us."

Christ answers: "It is not good for you to see the abyss. But if you
wish it, I will keep my promise. Come, follow me and see."

> And he led them to a place called Cherubim, that is, place of truth.
> And he beckoned to the angels of the West. And the earth was rolled
> up like a papyrus roll, and the abyss was exposed to their eyes. When
> the apostles saw it, they fell on their faces. But Jesus said to them:
> "Did I not say to you that it was not good for you to see the abyss?"
> And he again beckoned to the angels, and the abyss was covered up.

As soon as the angels have covered the abyss, Bartholomew asks to
see Satan. Jesus warns him "I tell you, when you see him, not only
you but the apostles with you, and Mary, will fall on your faces and
will be like the dead." They all reply, "Lord, we wish to see him." The
earth shakes, Satan, at this point called "Beliar," comes up from the
underworld, huge, with a face "like a lightning of fire, and his eyes like
sparks." "A stinking smoke" comes from his nostrils, while his mouth
is "like a cleft of rock" and his wings are eighty yards long. As soon as
the apostles see him, "they fell to the ground on their faces and became
like dead men." But Jesus revives them and allows them to interrogate
Satan. At the end of Satan's brilliant description of his personal history
and his present duties in Hell, Bartholomew asks Jesus, "Lord, may I
reveal these mysteries to every man?" Jesus replies, "Entrust them to all
who are faithful and can keep them for themselves. For there are some
who are worthy of them; but there are also others to whom they ought
not to be entrusted," and he lists the boasters, idolaters, seducers to
fornication who are not worthy of the secrets. Jesus adds, "The things

are also to be kept secret because of those who can not contain them. For all who can contain them shall have a share in them."

* *

There seems to be a greed, or a thirst, for certain kinds of knowledge. Told of Mary's Annunciation, the apostles yearn to see the abyss; having seen the abyss, barely a breath later they crave an interview with Satan. Even when they are warned that they will fall flat on their faces and become "like the dead," their instinct is to say, "Show us, show us." Of course, one safeguarding factor is that when it is a divinity who is showing us, we can expect that he will revive us afterwards if we need it.

Like any great teacher, divinities seem to be fair game: the minute we find them, we buttonhole them and ask to see their faces, or the secrets of Heaven or Hell. We yearn to see more. In some ways it seems as though the divinity is seen as a vehicle towards further knowledge of things beyond or beside or around the godhead. Partly this has to do with the dangers of looking at a god face on. But perhaps, too, there is the obstacle of the god's unknowability, an obstacle that the prophet and the mystic are always trying to get around. Along with unknowability comes ineffability: we talk only of the things we can actually talk of, see only the things we can see. Thus the painter of the astonishing illuminations of the Uigur manuscript of the *Miraculous Journey of Mahomet* can show only gold flames when Mohammed is in the presence of his god, and cannot bring himself to show the divinity. In contrast, Michelangelo, full of bravado and mischief, feels he can expose the divine "back parts," as well as the divine face from time to time.

There is something interesting going on with what Jesus says about whom one should tell the mysteries to: those who are worthy, and those who can contain them. I take "contain" to mean both "not be demolished by" and "not reveal inappropriately."

In Jewish tradition, in the Babylonian Talmud, there is an even stranger and more restrictive example of who may be told what. That is, you are not supposed to teach the definitions of incestuous marriages to more than two people at a time; you should not teach the story of Creation to more than one person at a time; and you should not teach the Vision of Ezekiel to *even* one person, unless that person is wise enough to already understand it on his own.[32]

Who is allowed to see or to understand? The one who is worthy, and also prepared; the one who understands to some degree already and so will know the worth of the vision and not be destroyed by it.

The Local Forbidden

Perhaps the more local or domestic counterpart of glimpses into heaven or hell can be found in the childhood act of spying in our parents' bedroom. When they are not there, I mean.

Why do we do this snooping? What are we furtively looking for? Even when it is not expressly forbidden, we do not want to be found doing it.

Part of it, of course, is simple mischief. In the parental bedroom, of course, we're looking for the paraphernalia and evidence of sex, even if we don't know that is what we are after. Beyond the condoms and obscure equipment and the manuals and the picture magazines lurk all the questions of what actually goes on in the dark, the generative mysteries. Sometimes what we find is unintelligible to us, and we don't make sense of it until much later, if ever. Then, too, there are questions we don't dare utter or know how to formulate. Perhaps we are looking not only for explanations but also for further mysteries.

❧ ❦

My father's bedroom had two mysteries. One was the collection of his diaries, which he had kept faithfully from the time he was twelve. The mystery was not their contents, for he encouraged us to read them, and we found them charmingly boring. The mystery was why he felt compelled, almost from the time he could form his alphabet, to report in writing each day on the weather. This took me several decades to understand.

The other mystery in my father's bedroom was a machine in a glass case that ticked softly, like an old clock, with an almost inaudible whirring burden underneath. On a mahogany base sat a stack of seven flat metal cylinders, and beside them a round drum covered with graph paper on which two pens marked wavering lines in black and purple ink.

My father was passionate about atmospheric pressure and its vicissitudes, and also about the temperature. But atmospheric pressure, as it cannot be seen or felt, falls in the category of things that must be looked at slant, and the beautiful mahogany and brass contraption—the barograph—is a simple mechanical way of translating the unseen into the visible. The pen with black ink records the barometric pressure onto the graph paper wrapped around the drum, while the pen with purple ink traces the temperature. The graph paper needs to be changed every week, and that is when you wind the gear mechanism as well. And if you have forgotten to do this in a while, then anticipation of dramatic weather events will remind you to set it turning once again.

If my forbidden looking in my father's room consisted of trying to see inside that barographic machine of forbidden looking, I was better at taking it all apart than putting it together again—unscrewing the nibs, unwinding the clock mechanism that drove the drum, fiddling with the seven cylinders, called *aneroids*, shaking them to see what was inside. I did not know then that there is nothing inside them, they are mostly vacuum, except for the springs that keep them from collapsing in on themselves, and that the sevenfold summation of their minute changes in volume due to changes in atmospheric pressure are what is being recorded by the black ink.

൭ ൴

My mother's room held more mysteries. Chinese ceramic vases and incense burners, glazed in oxblood or turquoise, seemed to be meant for greater treasures than the odd button, the paper clip, the folded index card. Then there was the bottom drawer in her bureau that held only purses and evening bags. These were all strange and had to be examined. Partly for the hints of what it meant to be adult, and partly because whenever she went to a wedding or other exotic place, I would line her purse with waxed paper, to encourage her to bring delicacies home to us. Perhaps some forgotten petits fours still lurked there. My interest in nondomestic foodstuffs began very early.

My mother's mirrored dressing table had something that puzzles me even today: a small glass vial containing infinitely small silver

Figure 15. "The American Cowslip" by Robert John Thornton (see Plate IV).

spoons. No, no, I mean much smaller than you are thinking. Perhaps you could fit five grains of fine salt on one of these spoons, but you could not pick it up except by licking a finger to dab it. Were they some form of primitive sparkles? Beauty marks? These spoons remain a mystery. I cannot find anyone who remembers them or knows what they were used for. I may have dreamt them.

A strange picture, framed in black wood with a black glass mat, used to hang in my mother's bedroom. This hand-colored engraving of the plant called Shooting Star, or *Dodecatheon meadia,* spooked us as children despite its reassuring label, "The American Cowslip" (Figure 15). Although it was not forbidden to us to look at, we did not dare to examine it too closely and when we mentioned it to each other we called it "The Evil Cowslip." Perhaps this was silly, for plants are probably beyond the reach of human morality—though my favorites often retain hints of darkness and death, intoxication or plunder.

In that engraving, stunted trees grope in front of caves filled with obscure menace. Beyond the crags the sky glowers over an unquiet ocean where two galleons chase each other with intense and unknown purpose. The flower, the Shooting Star, grows from a peculiarly barren ground, its single stalk rising tall from primrose-like leaves to terminate in an explosion of pink blossoms whose petals are flipped backwards. The whole plant looks otherworldly, like a sheaf of fireworks bursting in all directions. Perhaps it is not of the vegetable kingdom at all, but demonstrates some geometric proof having to do with the shape of the cosmos or the ideal forms. Or does it represent something even further from matter—some sort of abstract process?

Framed in black glass, this picture looked like a death announcement, and it seemed odd to me as a child that my mother would want to keep it so close to her, in the room where she slept. She did have other anomalies on her bedroom walls: among the very Bostonian watercolors was a dark etching of a mother and child by Käthe Kollwitz, a German Expressionist who documented widows and orphans, poor people, and the suffering of war. I did not understand until much later the importance of what one keeps in one's bedroom, that these walls are for the things we want to look at in private, and that "The American Cowslip" and the Kollwitz etching were telling her something that she felt she needed to hear.

When I learned that my mother was dying, the first question that pierced me was: But who will take care of her gardens? I asked her if she wanted me to weed, to plant, to prune. She answered cryptically, "Face facts!" She was a brilliant gardener. She taught me that gardens are what gives texture to light. And to breath. From her I learned all the necessary names, but we never talked about the engraving of the Shooting Star, and I never saw one growing in her garden.

After my mother's death, and then my father's, I inherited the black-framed engraving. It turns out to be a page—ripped—from *Temple of Flora,* published in 1801 by Dr. Robert John Thornton. And as for the

Shooting Star itself, I finally tracked it down. It is a breathtakingly tiny plant—complex, and beautiful, pink with hints of mauve, a band of yellow, dashes of maroon and jet, standing no higher than a handspan. It is an explosion of symmetries, a cluster of shuttlecocks for a toad. Whenever it appears in my local nursery, I buy as many pots as I can carry. And yet, as sweet as it is, the delicate Shooting Star still trails a hint of gigantic menace—because I first knew it through that gloomy engraving from that plundered and dismembered volume.

❧ ❦

As I write about the Shooting Star, my old seasonal craving has just started up again, for a wild lily just now in bud. This is the American Turk's Cap, *Lilium superbum*, which grows in the lee of the hedges along the country roads here in southeastern Massachusetts. It flourishes in part shadow, in ground at once disturbed and protected. I have been yearning for this lily for years, but whenever I ask my friends about it, no one seems to have seen it but me. Rising six, eight, or even nine feet tall, above whorls of gray-green leaves, the candelabra-like tiers of brilliant flowers with their recurved petals resemble a group of turbans peeking out of the foliage—as though Turkish mystics were hiding in the hedgerows. The color of their religious headgear sometimes seems to shade miraculously from purest red to purest yellow without passing through orange. A holy impossibility. This lily so besets me that I feel each particular plant deserves a name, a given name I mean, as an individual.

There is a stand of these American Turk's Caps at the far edge of a cornfield near our house, just where the hedge has an opening to let the tractor in. This field does not belong to us, so I plot to kidnap the Turks.

❧ ❦

Dawn. The corn has just sprouted. I set out across it with my clippers and long black rubber gloves against the poison ivy. Today I will just prepare the Turk's Caps to be carried off, by disentangling them from the undergrowth of the hedgerow.

I find six plants, enough to leave some for the elderly farmer who owns the field. It would bother me to abscond with all his lilies if he loves them. The only other time I have poached, from a stand of dog-tooth violet that I found during a kayak-and-trespass, my entire haul died completely. This does not bother me; we learn by practice.

In my haste to cut away the brambles I break one of the lily stalks. I am a murderer now, no longer a simple thief, and am stricken with shame. The lilies are taller than I am, and so fragile that I will need a stretcher to get them back to my garden after digging them up. Possibly a bed sheet wrapped around broomsticks would work. Though he is violently against thievery, I will try to get my husband to come with me and hold one end of the stretcher. I could convince him that it would be saving the plants, for the remaining stalks are not robust, and have few and misshapen buds. They clearly need my help.

When I break the second lily—really, you only have to touch them and they buckle—a low voice startles me.

"Boo!"

It is the old woman who spends her days walking up and down the road by the cornfield, wearing a bright neon vest as though she were a member of a road crew. She is, I think—but it is a different sort of road, and crew.

"Hey there!" I say, trapped in the act. "How are you doing?"

"What are you up to, then?" she says, as though settling in for a chat. Her hair hangs every which way from under her hat, and that itself is askew. She carries a plastic bag in the manner of the homeless, but we are out in the country, and this is her territory, not mine.

"Oh, I'm just cleaning out," I say. "Around these lilies."

"Of course you are," she says. "Dear."

"They need some air, some light." I try not to stutter.

She bobs her head once more, but stays right there, watching me. I cut back brambles, blackberries, lay them to one side.

"Of course, I've done this myself," she says, peering at my face, then adds as though to clarify, "Lilies."

She does not leave, and her inspection makes me bashful. Finally I have to ask, "Am I doing the right thing?" But when I look up she is gone.

The next few mornings are full of mist. For days I hover, dithering, hinging in and out of actually going to the lilies. Each morning I take my shovel and my pail and set off across the foot high corn, debating whether I am more likely to kill or to save; and then, each morning, too soon the mist lifts, I feel exposed, and I turn back.

When the corn grows too high to step over I decide to wait until the blooms come and fade, in case I break the rest of the stalks. I turn back to my own garden.

☙ ❧

Have I been fickle here, deserting my delicate Shooting Star for a while in order to talk of my rape of the Turk's Caps?

What is so odd about the Shooting Star is that its fascination is the reverse of the usual: instead of a floral trumpet with its forbidden cavern that we try to look into but must penetrate with our imagination until we get to its dark secret cove, this plant penetrates our space with its syringe like point and its flipped back petals. I have watched for hours, wanting to know how it gets from the bud, whose concave petals point forward in the normal way, enclosing and protecting the hidden sexual parts, to the mature flower with its everted shuttlecock form. But I have never been patient enough to catch what must be a kind of vegetal spring-mechanism in action.

Perhaps Dr. Thornton would know more about how this odd and lovely flower works. Thornton was a Public Lecturer on Medical Biology, and he knew the darkness of flowers as well as the light: many of the plates in his *Temple of Flora* are as melancholy and brooding as my "American Cowslip," and you would not want to venture into his dismal landscapes alone. Occasionally he goes wild in his descriptions as well, as when he takes on the *Arum Dracunculus* or Dragon Arum:

> This extremely foetid poisonous plant will not admit of sober description; let us therefore personify it. *She* comes peeping from her purple crest with mischief fraught: from her green covert projects a horrid spear of darkest jet, which she brandishes aloft: issuing from her nostrils flies a noisome vapor infecting the ambient air: her hundred arms are interspersed with white, as in the garments of the inquisition; and on her swollen trunk are observed the speckles of a mighty dragon: her sex is strangely intermingled with the opposite! Confusion dire!—all framed for horror: or kind to warn the traveler that her fruits are poison-berries, grateful to the sight but fatal to the taste, such is the plan of *Providence* and such *Her* wide resolves.

In comparison, Dr. Thornton's view of the Shooting Star is mild and gracious. He calls it "a vegetable sky-rocket in different periods of explosion," and "a number of light shuttlecocks, fluttering in the air," and notes Linnaeus called it *Dodecatheon* after the twelve Olympian Gods, on account of its singular beauty and the multiplicity of its flowers.

Thornton's rendering of the Shooting Star is meticulous in showing each blossom at a different stage of development, from the immature bud through full flower to final seed pod. Against the glowering background, the plant seems to radiate light as it demonstrates the cycle of life, decay, and rebirth from seed. Perhaps that is what my mother saw in this engraving that led her to keep it in her bedroom: a sense of that continuous cyclical progression, at once luminous and mortal, which resides in the inner consciousness of the gardener. This progression is often so smooth and continuous that we cannot see it happening unless we look away and look back, fragmenting time so we glimpse discrete steps of existence. Night performs this beautifully, this temporal partitioning of bloom; this is one of the manifold functions of night. Thus, each morning the first necessity is to inspect the garden and gauge how it has given new texture to time.

ॐ ॐ

If I oscillate quickly enough here, I can contain both the towering Turk's Cap and the diminutive Shooting Star. Beyond any notions of darkness, death, and poaching—beyond mortality or morality—what is important about these plants is how they anchor Light into Matter, giving to Light qualities that we can apprehend. Then, they let us see the shape of the invisible wind. They also offer us that skewed glance of imagery with which to look at two different aspects of the nature of thought. The Turk's Cap Lily, with its clarity of colors and its ordered tiers of blossoms, is a demonstration of the ascending progression of rational argument, while the Shooting Star—with its pointillist eruption of fireworks—is the explosion of revelation.

How Ovid Sees Orpheus

If Orpheus had *not* looked back, it is quite possible that Eurydice would have always held it against him, and their marriage would have been filled with the strife of doubts and recriminations. "You don't love me, really—you didn't even look back when we were coming up from Hades." Perhaps Ovid has a notion of this when he describes the moment just after Orpheus has looked back,

> And she now, who must die a second death,
> did not find fault with him, for what indeed
> could he be faulted for, but his constancy.[33]

This is a strong statement that Ovid is making: that Eurydice does *not* find fault with Orpheus for looking back, and may be an indication that the proof of his love is crucially important to her, perhaps as important as resurrection.

In a stupor of grief, Orpheus tries unsuccessfully to cross the river Styx to get to Hades a second time. For three years, Ovid tells us, the poet mourns, avoiding women, and inventing pederasty among the Thracians. Singing his poems in a meadow, Orpheus calls an entire grove of trees into being, and in their shade, surrounded by birds and beasts and followed even by stones, he sings—until the maenads, worshippers of Bacchus, find him. Furious that he has been rejecting their overtures for so long, they attack the poet with spears and stones. As long as these hurled objects can hear Orpheus's song, however, they are enchanted and remain harmless. But the fury of the maenads increases until their own cacophony—of flutes, horns, drums, hand-claps, ululations—drowns out the music of his lyre. Then the spell is broken, and "the stones were reddened with a poet's blood." Brandishing rakes and hoes and mattocks that panicked farmers have left behind, the maenads strike Orpheus down,

> and past those lips—ah, Jupiter!—to which
> the stones would listen and the beasts respond,
> his exhaled ghost receded on the winds.

The maenads dismember him. As his head and lyre float down the Hebrus River, the lyre continues its plangent music, and Orpheus's tongue still sings; the riverbanks moan in reply.

What Orpheus Gains

While Virgil stresses youth, greed, impetuousness, and disobedience, Ovid focuses on love. I think that Ovid often indicates that Orpheus is by no means the heedless youth that Virgil makes him out to be.

Virgil does show Orpheus as an enchanting singer, but Ovid goes much further and shows him as the greatest of poets, the most brilliant of musicians, able to reason with and convince the divinities of the Underworld to do something unheard of. For Ovid, Orpheus sings the trees into existence and not only do birds and beasts come to listen to him, but "stones skip in his wake."

In the end Orpheus may prove to be wiser than most heroes. Even when he looks back, I think he knows what he is doing.

In his argument to Hades and Persephone, Orpheus says: Look, you're going to get both of us here in Hades in a while, anyway. We all know that this is our final and eternal home. What does it matter if you have her a few days less or more? Why don't you just give her back to me to live in the world above for the very brief time before we come here?

Orpheus is fully aware of the relative time—momentary versus infinite—spent above and below. In fact, it may be the opposite of greed and impetuousness to do what he does, for by *looking* Orpheus is ensuring an infinite joy with his beloved, rather than the short-sighted *not-looking* that would have gained her momentarily, but always, during life as well as for the infinite afterlife, with the marital strife and blame of inconstancy: "You never once looked at me."

Finally, Ovid gives a lovely picture of what does happen when Orpheus, after his terrifying murder by the maenads, makes his final descent to the Land of the Dead. In the Elysian Fields, he finds Eurydice and passionately embraces her.

> Here now they walk together, side by side,
> or now he follows as she goes before,
> or he precedes, and she goes after him;
> and now there is no longer any danger
> when Orpheus looks on Eurydice.

Note how Ovid changes tense here, from past to continual present as he shows Orpheus and Eurydice in everlasting blissful dalliance.

Disobedience in General

Let us consider all the magnificent stories in which the protagonist, told not to look at something, in fact obeys and forbears to look.

Actually, I do not think there are any, not any grand ones.[34]

I think that probably what is gained from looking far outweighs what is lost, whatever the consequences, and the story of looking is only linked with losing in order to mask or veil the dangerous brilliance of seeing. Whatever loss is entailed, it may carry its opposite—unexpected, and rarely understood—within it, in the same way that absence implies not nonexistence, but existence, and therefore presence. Not doing calls forth the idea of the action. When we fast, it is not that we are *not* eating, but that we are not *eating*.

The reason that there are no great stories about the obedient refrainer from looking at the Forbidden is that the good that comes from this non-action is so meager and single-stranded. It is the easy and sure and foretold good of the accountant, adding a column of numbers correctly, as opposed to the chancy and complicated vision of the poet or the mathematician or the mystic.

ॐ ॐ

I should make it clear that I am not advocating random disregard and disobedience. Forbidden looking, and, in the primeval case, forbidden tasting, is different from other acts because of the metaphorical connection between seeing and understanding. To grasp with the eye is to grasp with the mind. Plato makes this connection, as do many of the ancient philosophers who see the act of vision as an active thing, involving beams coming from the eye to the seen object.[35] Anyway, here I am mainly concerned with seeing and understanding, rather than generalized transgression.

Except, perhaps, for certain forms of plant poaching. Although I said that I was primarily interested in looking and seeing as forbidden acts, I should mention that stealing the Turk's Cap lilies finally occurred the following summer. By this time, perhaps because I had pruned the brambles that were competing with them, the lily stalks were now robust and did not buckle or break when I unearthed their strange dumbbell-shaped bulbs. I put the bulb ends in my bucket and let the stems rest on my shoulder as I went home across the cornfield through the dawn mist. The elation I felt is hard to defend, and also hard to describe, for it felt both heroic and epiphanic, as though I had stumbled on a new form of love, and the days and weeks that followed were in some odd way blessed by this discovery.

The stolen Turk's Caps are thriving in my garden, and the original seven plants have propagated themselves and are now seventeen in number. I am thrilled that they flourish, though I do not think that I am entitled to have them for that reason, and I still think of them as my purloined lilies, and of myself, not without pride, as a poacher. This puzzles me, for I do not find any joy or excitement in stealing things. Your valuables are completely safe with me. Yet, while I avoid gardens and private property, I must admit that public or ambiguous lands continue to tempt me; it is only a particular thicket of poison ivy that is, for the moment, keeping me from my next mission.

Forbidden Looking and Writing

When looking at some taboo things, we cannot see them, even while we are looking, because we are not ready. After we have undergone preparation, they become possible to grasp. Or we become able to grasp them. One of the places we are forbidden to look, in both senses—both because we should not and because we are unable—is into the mind and soul of another person. This is one reason why futuristic ideas of implanted computer chips to tell us what is going on in someone else's mind are so completely spooky.

Anyway, it is forbidden to inhabit another and if we do, then we may be inhabiting two people at once, our self and our other self, and what we are doing seems awfully close to schizophrenia, or multiple personality disorder, or some other form of madness. It is considered sick, and may not be so far from the madness that is love.

How far into the mind of another we are allowed to see can depend on our training and initiation; psychoanalysts and psychiatrists and neurologists and priests all have uncommon entrance into the mind-workings of others.

But what is writing but taking up residence in the mind of another? I am thinking particularly of fiction, biography, and certain types of lyric and epic poetry. Here our specialized training involves empathy and imagination and experience. Of course, such residence is dangerous, for all the reasons that deep imagining is dangerous, among them that the imagined world and the imagined self become illicitly intense, more vivid than our own kitchens, or bedrooms.

What is actually forbidden about this seeing into another is that it breaks down the boundary between self and other, between self and self.

As, of course, does Eros. And as with Eros, there is bitter mixed in with the sweet, for our characters, too, can deny our importunings, when they resist us and do not let us see as far in as we want. We do not tend to get jealous of our characters, though we may be jealous of the characters of other writers, which shows that these feelings are really part of our relationship with the Muse. Our relationship with her can involve multiple layers of jealousy, and she can be as jealous of us as we are of her. Occasionally she seems to take pleasure in refusing our attentions and neglecting to visit for long stretches of time. If we forget ourselves or become at all careless of her love, she can engage in pouting and fearsome silences.

Does one of the forbidden aspects of Eros really have something to do with seeing into the hidden territory of the beloved? Why is it that we feel as we get closer to someone, mortal or divine, that we should be able to see even more? Moses comes to mind here, and Semele, and Psyche. Is such greediness always a part of love, or is it only present, as in these three cases, when doubts are rampant? Certainly Orpheus's looking back had a large component of doubt—he doubts Eurydice, that she will be able to follow him, that she will in fact follow him, though she never has cause to doubt him. Perhaps the gods of the Underworld felt that he would not be able to bear the complexity of her emotions—reluctance, longing, questioning, which he would understand the moment he looked back at her.

The injunction against looking is always coupled with an unspoken statement about the necessity of trust. The gods, both Greek and Hebrew, always try to make this decision: Trust *me* that I know what you should be doing, here, which is to say, *you* not looking. Trust me that I know how much or little it is best for you to know.

⌀ ⌀

When we fall in love, we get to know our beloved's hidden strengths as well as weaknesses. Weaknesses are often easy to make out, but some of the deeper strengths are more unexpected, such as new forms of generosity—emotional or erotic or temporal or confidential or intellectual. Is there any difference between a weakness and a strength at this level, or is a weakness like a weed, defined only as growing where it should not? There may be things one sees from the vantage point of love that one does not see from anywhere else: which weeds are to be treasured, which blooms have turned to weeds.

The Purpose of the Prohibition

DO NOT LOOK HERE. Do not read this. Do not look at me. Do not listen to this. Do not taste that apple.

Do these prohibitions have a purpose? It seems that they might, in fact, be necessary to the action. That is, that the obstacle embodied by that injunction—the climbing over the hurdle of prohibition—is what gives height, vantage point, illumination, to the act. The prohibition defines and calls the act into being, and it sets the hero and the object of his glance apart, on a different plane, where everything he may do has such intensity of being that a single glance is monumental, volcanic. The injunction calls the universe into being.

I have heard it argued, and do believe, that the expulsion from the garden of Eden represents not a Fall, but a Completion, as Adam and Eve are transformed from pure spirit to a mix of spirit and matter—a completion that was both expected and necessary if they were to become fully human. According to this interpretation, in the phrase from Genesis 3:24 "And He drove out the human and set up east of the garden of Eden the Cherubim and the flame of the whirling sword to guard the way to the tree of life," the flaming sword is there both to light the way for the exiting couple, and to signify the glory of the completed act.

It is not that the Edenic, or any other prohibition, is meaningless. They are there to keep most of us safe, most of the time. To have life chug along on the tracks of the normal so we can get things done. Until it is time to call the universe into being, or to give humans their full complement of matter. Of course, knowing when it is time to do such a thing is the tricky part. And this is what takes preparation, practice, initiation. Not just anyone can go down to Hades and expect to get out again.

☙ ❧

The ancient Greeks had a single word having to do with seeing and saying that seems to me to capture this transformation between the forbidden and the performed. Greg Nagy discusses this word in his writings on myth:

> *Muô* means "I have my eyes closed" or "I have my mouth closed" in everyday situations, but "I see in a special way" or "I say in a special

way" in ritual. The idea of special visualization and verbalization is also evident in two derivatives of *muô*: (1) *mustês* "one who is initiated" and (2) *mustêrion* "that into which one is initiated."[36]

What a wonderful hinging between normal and ritual worlds—in a single word—the smooth working of the hinge depending on one's spiritual preparation or initiation; the room where we spend most of our time thus being the everyday and the room beyond being the *mustêrion*, the mysteries. Until we are prepared, we do, and had better, keep our eyes shut. But afterwards, we can, if we desire, see the mysteries, and perhaps even try to tell of them.

IV | Hell and Hinges Revisited

Christt Harrowing Hell

It is time for me to reconsider the question posed in Chapter I about the hinges in the images of Christ harrowing Hell: Why are the artists often so careful to show the hardware? To get at this, I would like to focus on two main types of representations of this scene. In what I will call the Byzantine form, which includes Greek and Russian Orthodox, Christ stands in the middle, facing the viewer, often holding a tall cross-staff, symbolic both of his crucifixion and of his victory over Hell. On his left and right are symmetrically arranged crypts from which he pulls the Old Testament prophets he is saving. Below Christ and below the coffins is the dark hole of the Underworld, sometimes

Figure 16. *Anastasis: Christ Saving Adam, Eve,* Byzantine fresco in the Chora Church (Kariye Camii), Istanbul.

containing a figure of Satan bound in chains. In the Greek Orthodox mosaics and frescoes, this dark region is often scattered with broken hardware, the locks and hinges of the gates of Hell that Christ demolished (Figure 16).

The Russian Orthodox icons showing this scene can be much more complex, as in the *Descent into Limbo* by Dionysius, from the Ferapontov Monastery in Russia (Figure 17). Here Christ stands on the broken gates of Hell, which are arranged in the form of a cross. He pulls Adam and Eve out of their open crypts; Adam, in a green robe is on Christ's right, along with patriarchs and Prophets; Eve, draped in red, is on Christ's left with male and female ancestors and Prophets, among them John the Baptist with a large halo. Christ is surrounded by a glowing blue sphere of the heavens, which is filled by spiritual powers in the form of haloed angels. Each angel holds a scepter with the name of a specific virtue: Happiness, Love, Wisdom, Truth, Humility, Reason, Life, Purity, Joy, Benevolence, Resurrection. The angels also have long thin red spears, with which they pierce the demons in the dark cavern of Hell below Christ. The demons, too, are labeled, as they represent various conditions thought to be vices: Death, Decay, Grief, Despair, Hatred, Falling, Vileness, Unreason, Deformity, Strife, Pride, and Mourning. At the bottom of Hell, in the center, two haloed angels bind Satan in chains, while on either side of the dark abyss, perhaps slightly in front of it, are the white-robed figures of those who have already been resurrected.

Commentary on this particular icon notes that stunningly absent from among the demonic vices are Fornication, Gluttony, Avarice, and Vanity. Apparently the priestly patrons of the artist were particularly fond of these qualities, which they demonstrated in themselves, and preferred to have them omitted from representations of the world after death.[1]

This icon is an example, perhaps better than any of the other illustrations in this book, of the tendency for paintings or engravings to act as treatises on religious or philosophical themes. Thus the hinge that so provoked me, initially, when I first looked at Fra Angelico's fresco of Christ in Limbo, really does have a necessary presence in the painting. I had been thinking in terms of *spatial composition* and balance, but these may be anachronistic concerns, for Fra Angelico and other artists of the Renaissance and earlier seem to be instead composing in terms of *meaning*. I think it may make much more sense

Figure 17. *Descent into Limbo* by Dionysius (see Plate V).

to read the elements of the composition as being there for a purpose, sometimes overtly didactic, as in Dionysius's icon, sometimes more contemplative, as objects to meditate upon. And as in writing, the balance between showing and telling can vary greatly, with the textual labels for the virtues and vices making this Russian icon veer much more toward the telling.

Figure 18. *The Descent into Limbo* by Duccio di Buoninsegna (see Plate VI).

Whether Greek or Russian Orthodox, the Byzantine form of Christ in Limbo retains a heraldic structure, with Christ standing in the axis of symmetry of the picture. When he pulls the dead from coffins on either side of him at the same time, I am reminded of the archaic hero of Mesopotamia and Greece, who is often shown grasping two beasts or birds or monsters by the neck, or feet, though the monster that has been overcome in these Christian paintings is, of course, the more abstract beast of mortality.

The other main type of representation of Christ in Limbo is the one we find in the art of Western Europe. Here Christ usually appears in profile view, at the entrance to the cavern of Hell where he reaches toward a whole crowd of people, as in the Fra Angelico fresco in Figure 1. Instead of the slender barred Cross that appears in some Byzantine depictions, Christ holds a staff with a banner with a cross on it, signifying both crucifixion and victory.

Whichever form of this scene we are looking at, we are also seeing the artist's own commentary about matters religious and spiritual. This is one of the jobs of artists, and they may have more freedom to make such commentaries than theologians. We who are reading such pictures must often bring all our attention to bear, in order to get these visual treatises to unfold.

The Broken Door and Where Christ Stands

Although I have been discussing Christ as a hero who descends to the Land of the Dead, in these paintings and mosaics he never seems to really penetrate Hell but always stays at the edge or in a sort of antechamber, neither inside nor outside. This has to do with the nature of Limbo.

Limbo is being done away with by the Catholic Church, so I would like to make a little discourse on it before it slips away entirely.

The word comes from the ablative singular of the Latin *Limbus*, meaning "at the edge or border," and is taken to refer to "a region supposed to exist on the border of Hell as the abode of the Just who died before Christ's coming, and of unbaptized infants" (*Oxford English Dictionary*; first recorded use, 1377). So we are talking of the threshold of Hades, *Limbus infernum*.[2]

Christ beckons and touches the formerly damned. But he communes only with the damned nearest the door, and the whole population extends far back into the cavernous depths where he does not go. Occasionally he stands on Satan who lies vanquished on the ground, as in the brilliantly opulent tempera painting by Duccio (Figure 18). Christ is never out of sight of day, which shines on his back, even though sometimes he himself is also a source of light. Sometimes he is accompanied by a man carrying a cross, whose identity will become clear below.

Christ stays in his liminal space; he pulls the people to him, to cross over the threshold, the ones he saves. Is it correct to really include him among the heroes descending to the underworld? Or does he just go to the door to give a sacred beckoning? There does not seem to be much at stake: though dead, he is already triumphant; he is going to Heaven; he has nothing to lose.

Figure 19. *Christ's Descent into Limbo* by Jacopo Bellini (see Plate VII).

Perhaps his heroism has to do with coming into contact with terrifying mysteries and with populations of demons. Sometimes the landscape is barren, deserted wasteland, with the loin-clothed patriarchs kneeling and desperate, importuning from the mouth of their craggy cave, as in Jacopo Bellini's rendering (Figure 19).

Figure 20. Detail of *Christ's Descent into Limbo* by Jacopo Bellini

Generally, however, in the Italian paintings, Limbo is rather pristine. The cavern of Hell is tidy and the dead wear superbly clean robes; even the demons seem known or familiar and rather tame, a bit like pet monkeys—mammals, anyway, often with dark feathered angels' wings. These demons are too tall to be cuddly, but they are attractive and often sexy, these Italian devils. They are skittish and fleeing, but you might think of keeping one or two in the backyard with your pet goat, and letting them into your kitchen from time to time. Such demons animate Jacopo Bellini's painting, which otherwise feels spartan, more desert-like (Figure 20).

If we move north to the Germanic world of the Protestants, however, the edge of Hell turns chaotic and scary; the devils are nightmarish and oppressive, reptilian and monstrous; there is unbearable psychic turbulence.

In Dürer's *The Large Passion*, for example, a flying demon with a woman's face hovers at the top of the picture. Ram's horns sprout from her bumpy head, while bat's wings, pendulous breasts, and a sea serpent's tail make her more terrifying. As she flies, she blows on a suspiciously obscene trumpet (Figure 21).

Figure 21. Detail of *Harrowing of Hell* from *The Large Passion* by Albrecht Dürer (see Figure 3). (Photograph © 2010 Museum of Fine Arts, Boston.)

Figure 22. *Christ's Descent into Limbo* by Martin Schongauer (see Plate VIII).

In contrast to the shy and mischievous demons of the Italians, the Germanic demons are always viciously on the attack. Some of the most disturbing ones have faces on their bellies or on their rumps, as in Martin Schongauer's painting—where, devil's feces litter the ground (Figure 22).

Having smashed the door, Christ usually doesn't put his feet on the ground. In Byzantine mosaics and Russian Orthodox icons, he hovers above or stands on the tall narrow doors of Hell that are sometimes arranged so as to remind us that the Crucifixion was necessary for his passage there.

As we move westward, to Italy and Alsace, if Christ is not hovering, as in the fresco of Fra Angelico in Figure 1, or standing on a devil as in the Duccio panel in Figure 18, he often stands on the massive door

Figure 23. *The Harrowing of Hell* by Martin Maler (see Plate IX).

he has torn down. This huge door acts as a ramp, a runway, a sacred teeter-totter or demon-swatter. No matter how many demons and devils may lurk in the cavern, one Satan can often be found under this fallen door, squashed like a cockroach as in Figure 1, or, as shown by the Protestants, not-quite-vanquished, as in the Schongauer painting.

In Scandinavia, England, and France, the entrance to Hell often appears as a monster's mouth. In the graceful fresco painted in 1425 by the Danish painter Martin Maler, the bestial Hell-mouth is lined by teeth and with fangs at top and bottom. Christ, with stigmata on his feet, holding his victorious flag, stands just outside the jaws of hell, grasping Adam with his right hand. Behind Adam are two more figures, one, in a long robe, might be Eve; the other looks like a young man, wearing a crown. Deeper in the maw lurk six humorous round-eyed demons with clownish red mouths. In front of the demons, looking a bit like a jester, sits a horned Satan, with reddish face and ochre body, and with ropes around his neck and body (Figure 23).

Nicodemus on Hell

Christ's descent to Hell does not appear at all in the Bible, although it is mentioned fleetingly in two of the four main Christian

Creeds or professions of faith.[3] The scriptural source for the artworks I have been discussing is probably the rich description in the apocryphal Gospel of Nicodemus. The earliest parts of this gospel appear in Greek dating to around 425 AD, but the narrative of Christ's descent appears to be a later addition.[4]

There are many books of the Apocrypha, some Jewish and some Christian; some of these are included in normal bibles, others, like the gospel attributed to Nicodemus, are not. In general, the books of the Apocrypha are considered so transcendent and esoteric that they must be kept secret, or so secondary that they are of questionable value. (The word comes from the Greek verb *apokryptein* meaning "to hide away.")

In Section IV of the Gospel of Nicodemus, two dead men, the sons of Symeon, among those freed from Hades by Christ, write down identical versions of the descent, in which they account a conversation they had overheard between Satan and Hell; Hell is personified as "Hades (god)" in this account. Satan, warning Hades about Jesus, says:

> He did me much mischief in the world above while he lived among mortal men. For wherever he found my servants, he cast them out, and all those whom I had made to be maimed or blind or lame or leprous he healed with only a word, and many whom I had made ready to be buried he also with only a word made alive again.

Hades replies to the "heir of darkness, son of perdition, devil" that a short while ago he had swallowed "a certain dead man called Lazarus," but that someone forcibly snatched away this dead man "with only a word." Hades presumes that that someone is Christ and, if they let Christ into the domain of Hell, that he will take all the other dead as well. Hades continues:

> For, behold, I see that all those whom I have swallowed up from the beginning of the world are disquieted. I have pain in the stomach. Lazarus who was snatched from me before seems to me no good sign. For not like a dead man, but like an eagle he flew away from me, so quickly did the earth cast him out.

In one final howl Hades adjures Satan not to bring Christ there, because,

By the darkness which surrounds us, if you bring him here, none of the dead will be left for me.

"I have pain in the stomach . . . none of the dead will be left for me." Hades here is not only the *God* of the Dead, but also the *Realm* of the Dead, perhaps in the form of the Monster Who Has Eaten the Dead, as seen in Figure 23. He is definitely not the Hades known to the Romans as Pluto, the raptor husband of Persephone. He is the whole damned Realm and—he is telling us how he feels.

∂◦ ◦◦

For writers, the wonder of this passage is this: in all the other stories of the hero going to the Underworld, the narrator identifies with the descending hero and/or the dead whom he tries to save. Nicodemus is the only one I know of who makes the imaginative leap of thinking as Hades himself. Thus, except for this case, we never get to hear from the Land, the Realm, the bellyaching container of all the dead. What a wonderfully odd narrative.

∂◦ ◦◦

The Gospel of Nicodemus continues with the thunderous voice of Christ interrupting the conversation between Satan and Hades. Hades sends Satan out to try to withstand the interloper, and then tells all his demons:

> Make fast well and strongly the gates of brass and the bars of iron, and hold my locks, and stand upright and watch every point. For if he comes in, woe will seize us.

In one of the early versions of this gospel (the second Latin recension), right before Christ enters Hades, a man carrying a cross comes in. This is the "Good Thief," one of the two thieves who had been crucified along with Christ, and the one to whom he says "Truly, truly, today, I say to you, you shall be with me in Paradise" (Luke 23:43). The cross that he carries will be erected in Hell by Christ, to signify his victory. This robber with his cross occurs prominently in the painting by Jacopo Bellini in Figure 19.

When Christ finally attacks,

> The gates of brass were broken in pieces and the bars of iron were crushed and all the dead who were bound were loosed from their chains.

Here at last we may have the source for all the locks and bolts and nails and keys floating below Christ in Byzantine mosaics and frescoes, as in Figure 16.

Finally Christ enters into Hell, illuminating "all the dark places of Hades" by his presence. Seizing Satan, he hands him to the angels, saying "Bind with iron fetters his hands and his feet and his neck and his mouth." We see this elaborately bound Satan in the Byzantine fresco (Figure 16), in the Russian icon (Figure 17), and perhaps even in the Danish fresco (Figure 23).

The Importance of Hinges

The narrative of Christ's descent to Hell in the Gospel of Nicodemus explains the odd presence of so much hardware scattered in the Byzantine mosaics and frescoes of this scene. This narrative also accounts, in part at least, for the broken hinges of the smashed doors of Hell in Western European art. But I keep wondering if there might be more significance to these hinges and the way they are shown.

What if we swing back, again, to the notion of hinges and axles. An axle, we should remember, allows complete rotation around a still center. A hinge is the restricted version, allowing less of a rotation. All sorts of crucial things can be thought of as axles, or hinges, or that special and dangerous case: the gravity-stricken hinge that is found on a trapdoor.

Hinges in the Body

Let us begin really close to home: Start with the moment you get up in the morning. Perhaps you are old enough to have achieved a certain creakiness as the day begins, stiffness of the simple hinges of the knee joint, the elbow, the fingers. And what about the ball-and-socket joints of the hip and shoulder? Contrary to the action of the usual hinge, in the case of the ball joint it is the inner part that rotates, while the socket of the hip or shoulder stays still. Often with the ball joint there is no single axis of rotation. Even so, there tends to be this turning around, or within, stillness.

The joints of the body were crucial also to the ancients: think of all those limping gods, think of all the prophets and heroes with leg injuries. As a sign of his divinity, Pythagoras had a golden thigh, but could it also have been an early form of knee replacement? Pan, the dancing goat god, so unstable on his cloven hooves—is it just that he is up on his hind feet, or is there a limp as well? Achilles has a fatal vulnerability of the heel tendon. I am purposefully leaving out Hephaistos here, as there are occasional medical articles that cast doubt on joint problems as the cause of his limp. But do not forget that the Angel of God who wrestles all night with Jacob can only hold his own by finally dislocating Jacob's hip socket.[5]

We seem to have some inborn notion of what the normal joint looks like and how it works. "Double-jointedness" often looks wrong

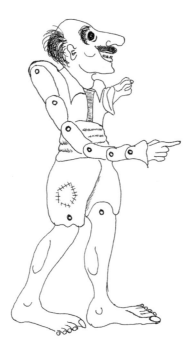

Figure 24. The Greek shadow puppet, Karagiozi, after a figure by the contemporary master puppeteer, Manthos Athinaios.

or strange, even though it is merely a sign of a single joint that is more mobile than usual; certain forms of Indian or Siamese dancing—in which the hands or feet seem to bend impossibly far—can be weird and alluring as well as fascinating and worrisome.

There is even a sort of humorous horror to the legendary Greek shadow puppet, Karagiozi, whose name comes from the Turkish for "Black eyes." This mustachioed figure with an impressive nose also has a hunched back. What is most uncanny about him, though, is his elongated many-elbowed arm. At once phallic and magical, this arm is superbly suited to his habits of petty thievery and erotic mischief. (See Figure 24.)

Closer to home, and even more uncannily jointed are spiders. Is it not their very jointedness that inspires some humans with love, and others with such intense phobia? Spiders' legs have seven segments while insect legs have only five.

Too much hinging of limbs can inspire us with awe or even fright. So, too, in the mind: imagery, meaning, or moods that flutter too rapidly between one thing and another can feel at times like poetic genius, at times like madness.

The Book as a Hinged Object

Leaving the hinges of our own bodies, let us turn to the physical object of the book. The book is full of hinges, with the covers and pages each allowing for oscillation and indecision as well as forward motion—by increments of a single page, or in huge chunks. The spine of a book acts as multi-door hinge, with that familiar limited range of rotation, and the action of opening a book mirrors that of opening a door, both physically and metaphorically.

In the beginning, the Hebrew Bible was a collection of scrolls, but by the middle of the fourth century, it was common for the New Testament to be in one large codex or book. Book technology was shifting, and some scholars think that the early Christians preferred the book form to differentiate themselves from the Jews with their scrolls.[6] In any case, once you separate out the parts you consider dangerous or spurious, and segregate them in apocryphas, the remaining canonical text or scroll or book becomes much more portable, and in general easier to deal with.

The Hinges of Heaven

Even further away from the articulations of our own bodies, consider the motion of the heavenly spheres as they orbit around each other and spin on their own axes. If, as the sun is setting, you turn your back on all that obvious glory and watch the eastern sky instead, you can see a more subtle sign of the earth's rotation—especially when the air is clear and the view long and unobstructed. Then a dark bluish-grey band appears on the eastern horizon. This is the shadow of the earth, projected onto the atmosphere.[7] Above the dark band there is often a brighter pink area of scattered light, known as the "Belt of Venus" or, more technically, the anti-twilight arch. As the sun sinks further below the western horizon, the dark earth shadow keeps rising in the east and the Belt of Venus dims until finally the earth's shadow is indistinguishable from the blue of the darkening vault overhead. On certain clear nights when this rising of earth's shadow is particularly visible, it can feel as though the vast hinged dome of night has risen up to cover our whole portion of the globe.

The fiery radiance of sunset is so compelling that we tend to ignore the rise of the earth's shadow. Our language seems to forget all about it, even though it reflects the intensity with which changes in light strike us: we say, Day breaks, as though the dome of night is shattered, and, Night falls, as though we feel suddenly mantled with darkness.

Like other momentous hinges, the diurnal ones—dawn and dusk—are visually noisy, as though light's increased labor, passing obliquely through the atmosphere, deserves spectrally raucous and triumphant proclamation. The hinges of the day are sonically noisy, too, full of the sudden sounds of birds at first light and again at twilight, when the awakenings and preparations of other night animals are heard as well, and bats erupt from their caves in a spume-like particulate smoke.[8]

Perhaps because of the shift in the rate at which the light is changing, twilight can bring a form of terror. At midday if we gaze on something for ten or twenty minutes, we do not usually notice great changes in illumination, but at dusk, the obliqueness of the sun's angle means that its light has to travel through progressively thicker atmosphere, and with each minute or two we can see the darkening process itself, everywhere, all around us, not just on the east and west horizons. As though Time is suddenly visible. Quickening.

Hesperian Depression

One of the most beautiful descriptions of odd things that happen at twilight appears in a book called *The Soul of the Ape*, by the South African writer Eugene Marais (1871–1936). This book was finished in 1919 but not published until 1969.

Marais claims that our moods are best at sunrise or in the early morning, but around sunset what he calls Hesperian depression sets in. (*Hesperus* is from the Greek for *evening*, and is another name of the evening star, the planet Venus.) In the great cities, he says, we cannot see or measure this evening melancholy because of all the city lights and amusements, all the ways we have devised to counteract what he calls "The Diurnal Crisis." He describes how in a native South African village he frequents, an air of silence and mournful dejection descends on everyone at sunset. After night falls, however, the village comes alive again with music and dancing and fires.[9]

And then comes Marais' most astonishing description of Hesperian depression in the Chacma Baboon (also called the Savannah Baboon). He tells how these baboons would collect in groups near their sleeping place before dusk, where the older animals would "talk," and the younger ones engage in romping games and mating, with the most joyous behavior occurring during the hour before sunset. But at sunset the scene transforms into quiet stillness, the smallest animals finding protection in the arms of their mothers, the playful ones gathering to watch the sunset. Now,

> The older ones assumed attitudes of profound dejection, and for long intervals the silence would be unbroken except for the soft whimpering complaints of the little ones and the consoling gurgling of the mothers. And then from all sides would come the sound of mourning, a sound never uttered otherwise than on occasion of great sorrow—of death or parting.

When darkness has settled, Marais says, the "evening melancholy" disappears, and the games are resumed and on moonlit nights might continue for several hours.

Although I do not always think of it as Hesperian *depression*, exactly, for sometimes the feeling of excitement at the coming of night is stronger than the sadness of the going of day, the descriptions in Marais' book underline the significance of twilight. That our close relatives among the primates also feel this odd hinging disturbance in the luminosity makes me wonder how widely spread in the animal kingdom this perception extends. Is it that at twilight there seems to be a slight hiccough in the smoothly turning diurnal rotation? Is this also what we are afraid of, not just the planetary motion, but that it won't all go smoothly? Or does the hiccough simply remind us of the rotation and its terrors? The primal centrifugal fear that gravity might forget to hold and let us spin off into the night. Or the mortal reminder that we are but featherless walking timepieces, and the earth we walk on is a clock, rotating within other vaster clocks, systems upon systems, all enmeshed like gears.

The Twilight of Creation

Twilight can be a time of unusual moods and behaviors for apes and humans. So too for the Hebrew God, especially on the sixth day of Creation.

For Jews, dusk is a particular time of doubt and uncertainty, over and above the universal shakiness due to the light changing so precipitously. This stems from the fact that the Jewish day begins not with the dawn, but with the evening before.

So day begins with evening, and evening is defined as when the sky is so dark that three stars are visible. Not comets or shooting stars, but three ordinary stars of medium size. But what about before that? What about after sunset and before we can see those three stars. What day are we in then? Is it yesterday or tomorrow? It clearly is not today, for the sun has gone below the horizon. We are on an edge at which today seems to have disappeared. This is the period called *Beyn Hashmoshes*, "Between the Settings."

Dusk on the sixth day of Creation is an extraordinarily strange time; it is a sort of triple hinge. On the first level, the week of Creation itself is the monumental hinge of the universe: on one side there is nothing except God—he has not even made angels yet—and on the other side, at the end of the week, there is everything.

The sixth day, too, is the hinge of the week, for it brings to an end the creative process as God puts in place the beasts and the cattle and the creeping things and man and woman and herbs and fruit trees, and sees that everything is very good. Everything is done, everything but the resting, which will come on the seventh day.

Sunset comes, and now there is that sixth dusk, hinge of the day, that time of doubt and uncertainty, when the light is mostly gone, taking with it the today of the sixth day.

Not in the Bible, but in Jewish legends (The Aggadah), we see that God continues working in this twilight before the first Sabbath.[10] This is when he makes certain peculiar magical, miraculous things that do not have a place in the ordinary Creation. Not many does he make, then, just ten or thirteen, depending on which rabbi you listen to. Among the things God makes at this time are the Tablets given to Moses, the Writing on the Tablets, and the Shape of the Letters of the writing (those are three different things). Some say he makes the Ram of Abraham that will take Isaac's place, and also the destructive forces that afflict mankind. But all agree that he makes the staff of Moses, that magical wand that, on demand, can turn into a serpent and eat other serpents made from the wands of Pharaoh's own, less powerful, magicians. In this twilight, too, God makes the Shamir, a mythical creature who can engrave diamonds but whose main characteristic is that he can cleave the stones for the construction of the Temple. Iron should never be used for the Temple, since iron, because it is used for weapons, works to shorten life. This diamond-cutting, stone-splitting Shamir is said to have the form of a worm and to be the size of a grain of barley.[11]

And finally some say that this is when God makes the Original Tongs, the *fundamental* tongs, for tongs, those hinged claws or pincers, can only be made on a blacksmith's forge by using other, previously forged tongs. So, if one goes all the way back, there has to have been a first, original set. To my way of thinking, these tongs might also have been used to fabricate the fundamental hinges of Hell, as though crucial hinges connect with each other in complicated ways. Thus the tongs constructed on the eve of the end of the week of Creation, which is itself a hinge of the manifestation of the universe, are used to forge a set of fundamental hinges for the gates between the realms of the living and the dead.

Hinges and Writing

Hinges in Poems

In poetry the most obvious hinge is the line break, or, as James Longenbach more correctly calls it, the "line ending."[12] Each line of a poem can have the fullness of a room in a house, and moving on to the next line can feel like stepping to an adjacent room. I said earlier that the word *stanza* comes from the Italian word for *room*, but I will hinge to another usage here, for I really prefer to think of the poetic *line* as more akin to a room, thus shifting the image of the poetic stanza to that of a house, or of one floor of a multi-story house. This is allowable, I think, because the Italian word *stanza* means "dwelling" or "abode" or "stopping place" or even "room," from the Latin *stare*, "to stand."

When there is a strong disjunction between two successive lines, when they seem to be talking about completely different things, their physical proximity on the page allows for a sort of shimmer between at-first-glance unrelated images or thoughts, their juxtaposition hinting at, or bringing about, some relation between them, as though one could pivot back and forth on a swinging door of language.

Ambiguities of syntax or meaning perform a hinging function, too: setting up an oscillation that can become so rapid that one ends up in two places at once. Logically speaking, the hinge of ambiguity can act as the inclusive *or*. Is Light a Wave? Yes. Or a Particle? Yes.

The turn of a poem can feel like a spiraling upwards to a new vantage point, often executed so smoothly that it is only much later that one realizes the new elevation. But sometimes the turn hinges on a more discrete shimmer, or even a linked series of oscillations.

Then, too, there are poems that suddenly use the second person, abruptly calling the reader, and this introduces a wonderfully discrete trapdoor, a chute leading to a different realm of existence—for once the speaker addresses us as "you," this carries us into the poem, we become not only listener/observer, but also participant—now, with the speaker, on his little boat, transported.

Hinges in Fiction

In fiction, the most obvious formal hinge is the section break, allowing contiguous parts of the text to veer off from each other in time or space or viewpoint. Paragraphs and chapters can work in a similar fashion.

But if we shift from form to content, an important pivotal element is the small detail—like the pintle hinge on the doorframe in the fresco of Fra Angelico, or the hinge on the smashed wooden planks in Dürer's engravings. The broken hinges in those pictures function as a visual alarm, pictorially announcing the crossing of a threshold, a crucial change of state. It makes sense to look for such things in fiction, the odd sensory details that seem superfluous but work to ground us in the importance and credibility of matter and pierce us with intensity of feeling. "Away Laura flew, still holding her piece of bread-and-butter," or, "Beside the gate an old, old woman with a crutch sat in a chair, watching. She had her feet on a newspaper." Mansfield's "The Garden Party" is full of such details, seemingly homey and ordinary but sharp edged and piercing as they work to announce a threshold of being.

Hinges and Trapdoors as Plot Elements

Remember the difference between the hinged door and the trapdoor hidden in the grass: the trapdoor feels unidirectional. You plunge down; then, in order to get out again, you must struggle wildly to regain ground level. Greg Nagy pointed out to me one of the earliest and clearest examples of the trapdoor in Greek mythology, occurring in the Homeric "Hymn to Demeter." Persephone is gathering flowers in the meadow with her friends when she sees a magical narcissus with a hundred blossoms on one stem:

> Amazed she stretched out both hands to pick
> the charming bloom—and a chasm opened
> in the Nysian plain. Out sprang Lord of the Dead,
> god of many names, on his immortal horses.
> Snatching the unwilling girl, he carried her off
> In his golden chariot, as she cried and screamed aloud. . . .[13]

We know we are in a special realm when someone picks a flower using both hands—you cannot break the stem that way because you cannot get the necessary sideways torque, as both hands counteract each other. This is not how you pick a flower—but it is how you pull something up by the bulb. And when you do, the chasm opens. Later in the same poem, when Persephone is telling Demeter what happened, she says of the narcissus:

> When I picked it in delight, the earth gave way from beneath,
> And the mighty Lord of the Many Dead sprang out.[14]

Notice how Persephone picks her words here, with the ground giving way beneath her feet. This feels perilously close to a description of what it feels like to fall in love. I can never keep from wondering if perhaps she is protecting herself from her mother's wrath when she adds:

> Hades dragged me most unwilling under the earth
> In his golden chariot; I shouted and screamed aloud.

Demeter is worried, panicky, furious about her daughter's abduction. Eventually, she goes to Mount Olympus and works out an arrangement with the gods, and Persephone will spend only one third of the year in the kingdom of Hades, away from her mother; Demeter brings winter to the earth then, to give us our own measure of her personal depression and gloom.

I know this is an abduction story. But it is also the age-old story of the adolescent girl causing grief by running off with the wrong man, a swarthy stranger from a different land. Kidnapping, yes. But we should remember what Persephone had been doing when this happened: poaching plants, pulling up the magic narcissus with its bulb. Before being abducted, she abducts. In some sense she also abducts Hades, seducing him, causing him to carry her off.

In fact, in ancient art works, Persephone does not look like a victim, nor does she seem at all unhappy to be in the Underworld. Instead she looks regal, mysterious, knowing, and erotic; the erotic is perpetually linked in various ways with the Underworld.

It is no wonder that this tale should have at its center the daughter of Demeter, goddess of agriculture and fruitfulness. Poaching, kidnapping, rape, plunder—all of these are crucial to gardens. Birds and field mice and things that creep along the ground all scatter seeds by absconding with them, carrying them from place to place. And humans, when they farm, what do they do but take the seeds from spent flower heads, and sow them somewhere else of their own choosing, or bury them in the earth.

Passively covered by earth, actively extending roots into the soil, each seed spends necessary time in the underworld. Perhaps there is some connection between planting rituals and chthonic religions. In any case, I think we are wrong if we see Persephone as victim, or if we view her only as that, for she rules and determines the underground habits not only of the dead, but of all growing generative things as well—their incubation, rootedness, and eventual propagation. She takes care of dormant ideas until they sprout.

The Hinge of Empathy

I would like to turn back to that conversation between Satan and Hades in the Gospel of Nicodemus. The complaint of Hades, "I see that all those whom I have swallowed up from the beginning of the world are disquieted. I have pain in the stomach," brings us into the mind and body of the most monstrous realm we can think of, and, for a moment at least, we have a glimmer of what he feels. What a brilliant turn, to speak in the voice of Hell himself. How wonderfully this demonstrates the hinge of imagining—the empathic shift into the consciousness of another that is at the core of writing. This shift, this intense and total inhabiting of one's characters, feeling their feelings, is necessary and terrifying. Without such an inhabiting one risks remaining distant and judgmental; one is, at best, compassionate. Compassion is a state that always feels a bit safe to me, because it is contrary-to-fact. The compassionate statement, "Oh, the poor thing!" has within it, I think, the implication, "Oh, how awful I would feel if that were to happen to me . . . but it hasn't."

The empathic counterpart is, "Oh, it is happening to me, and it is awful." And here lies the danger, for if one is really inhabiting the characters one creates and writes into existence, then one's emotional load is intense and one's situation borders on schizophrenia. One is, at the same time, oneself and the other, the writer and the written. And so one oscillates between these two states, these two psyches, touching the solid ground of self, or trying to, when one feels the need, or when the terror is too great.

Occasionally someone asks me why the people in my fictions are never punished for their misdeeds, or at least, not by me. This question always takes me aback. I babble and stutter for a while, as I try to explain that my responsibility toward my characters is to feel their feelings, to be a mouthpiece for their voices, to show how they respond to their predicaments and rub up against the world.

"Music from Spain" and Its Trapdoors

For an example of how hinges and trapdoors work within the plot of a piece of contemporary fiction, I would like to turn to Eudora Welty's long short story "Music from Spain," which is a chapter in her brilliant story cycle, *The Golden Apples*. This tale of a descent to the

world of the dead is framed by double trapdoors; there are also crucial hinge points throughout, where the two main characters oscillate
between positions of helplessness and power.

One morning at breakfast, Eugene MacClain's wife tells him he
has a crumb on his cheek—and he slaps her face. This is the first trapdoor, for Eugene has never done such a thing. He cannot conceive of
such an act. All he can think to do is to rush down the stairs and out
of the house and away, before Emma can stop him, before she can call
his name into the outside air.

Eugene finds himself a stranger, in both senses: he no longer recognizes himself, and he finds a stranger to spend the day with.

The outside world, once he steps into it, has become unfamiliar,
and Eugene needs this uncanniness, just as he needs a stranger, to
help him to see and utter what he must. Although it is morning in
San Francisco, for Eugene it has become Nighttown—that uncanny
cosmopolitan analogue of the world of the dead. In Nighttown, as in
Hades, the rules of behavior are suddenly bendable, and boundaries
between things—genders, species, realms—become indistinct. This
blurring of boundaries is one of the surest signs that one has entered
the other world. To Eugene, the eucalyptus trees look like birds fluffed
up in the cold, and he feels that he can hear the beating of their arbo-
real hearts; he himself jerks like a pigeon, while his feet stamp on the
leaves like hooves; the sun and moon seem to have changed places, the
sun "smaller than the moon, rolled like a little wheel through the fog"
and perhaps more weirdly, actions, such as striking one's wife, turn
into animate beings:

> His *act*—with that proving it had been part of him, turned around
> and looked at him in the form of a question. At Sacramento Street it
> skirted through traffic beside him in sudden dependency, almost like
> a comedian pretending to be an old man.

In spite of the fog, the day sparkles; everything is transformed, spec-
tacular, transcendent. As the bloody carcasses of dead cows are wheeled
across his path to the butcher shop, Eugene realizes that he will not go
to his job as a watch repairman at Bertsinger's Jewelers, "today he was
not able to take those watches apart." Broken watches thus will not get
fixed on this day, and we know that anomalies or distortions of time
are often a sign of the other world.

In fact, time is a crucial element of this story. It is a broken-hearted time, exactly a year after the death from pneumonia of Fan, Eugene and Emma's little daughter. Fan's presence sprinkles the narrative. As Eugene passes the fish market, everything is laid out as though for a holiday: salmon steaks arranged in a "fan" shape, filets of sole displayed in a chevron pattern, reminding us, and Eugene, of little Fan's blond braids.

As Eugene walks through the city, he keeps questioning himself as to why he slapped his wife, Emma, and why now? Twice Eugene thinks that the slap had been "like kissing the cheek of the dead," as though we are to take it in two ways: Emma is still so numb with grief that she cannot react even when her husband slaps her for the first time; and, by this act, Eugene will try to bring to life, or at least come into contact with, the love and grief of which he has been unable to speak.

Eugene goes down Market Street, past shops of prosthetic devices and posters for the medium, Madame Blavatsky, and finally recognizes someone he does not know—a Spaniard, dressed all in black with long black hair and a large black hat. Eugene hurries after him, proud of having recognized this "stranger but not a stranger," for this is the guitarist he and his wife Emma saw in concert the previous night. When the stranger is just about to step into the path of an oncoming car, Eugene flings himself forward to grab him, pulling him back, saving his life. This is Eugene's second trapdoor of the morning. He and the Spaniard stand on the curb, shaking hands, until Eugene realizes that the Spaniard speaks no English; however, the momentousness of having saved his life means that Eugene cannot just walk away:

> Then the order of things seemed to be that the two men should stroll on together down the street. That came out of the very helplessness of not being able to speak—to thank or to deprecate.

The difficulty or uselessness of spoken language is often a marker that one is in an otherworldly domain.

At lunchtime, Eugene leads the Spaniard to a fancy restaurant, and here the stranger seems oddly at home, talking French with the waiters, and ordering an immense meal (perhaps a paella?) that includes clams and pieces of chicken, whose bones he seems to be spitting out for the rest of the afternoon, like some ogre.

While walking after lunch, the two men witness the death of a woman hit by a trolley. They explore near the old Spanish graveyards, and again the Spaniard is so nimble, so sure-footed, it is as though he is at home or has been there before.

This Spaniard, who is he? He smells of smoke and travel, he is aloof and ravenous, the muscles of his face group themselves with "hideous luxuriousness" as his grape-colored lips surround a sweet-smelling cigarette. At one point Eugene visualizes him:

> turning his back, with his voluminous coat-tails sailing, and his feet off the ground, floating bird-like into the pin-point distance.

At various points in the story, the Spaniard's face has all the complexity of expression of some dark god: solicitude, meditation, amusement, sleepiness, implacability. At times he seems more satanic:

> the Spaniard with horns on his head—waiting—or advancing! And always the one, dark face, though momentarily fire from his nostrils brimmed over, with that veritable *waste* of life!

The two men walk along the beach and finally ascend to the cliffs at Land's End, the end of the world, where the Spaniard offers Eugene a Mariposa lily, and Eugene, who has been muttering questions and accusations to himself all day, says, "You assaulted your wife." I see Eugene's self-accusations and questions to himself as his form of the Traveler's Questions: because he and the Spaniard have no common language, Eugene has to keep asking himself who he is, how he got to where he finds himself, and why.

The Spaniard smiles and keeps presenting the lily. "But in your heart," Eugene continues, but stops when he recognizes that he can never express himself when it counts. Then he does an inexplicable thing: he grabs the Spaniard around his vast waist and realizes that the Spaniard is so light on his feet that he could easily throw him down the cliff to his death:

> he had only to make one move more, to unsettle that weight and let it go. . . . One more move and the man would go too, drop out of sight. He would go down below and it took only a touch.

As Eugene clings to the Spaniard, still toying with the idea of pushing him over the cliff, the Spaniard closes his eyes and bellows:

> Then a bullish roar opened out of him. He wagged his enormous head. What seemed to be utterances of the wildest order came from the wide mouth, together with dinner's old reek. Eugene half expected more bones. . . . The Spaniard's eyes also were open to the widest, and his nostrils had the hairs raised erect in them.

Eventually the Spaniard gets the upper hand and takes hold of Eugene, with "hard, callused fingers like prongs," and Eugene comes under the spell of a strange sensation, which he has felt before when very tired, lying in bed with his wife Emma asleep beside him:

> His mouth received and was explored by some immensity. It became more and more immense while he waited. . . . Only the finest, frailest thread of his own body seemed to exist, in order to provide the mouth. He seemed to have the world on his tongue. And it had no taste—only size.

Aside from the strong feeling of erotic dislocation in this passage, there is also philosophical astonishment that something as small as the human mouth or brain can perceive the cosmos or imagine the infinite. It is telling that Eugene can have such visions both while being held at the edge of the precipice and while in the cozy domesticity of being in bed with his wife. And although he has had this vision before, it feels as if it is being seen most clearly now, at this moment, *in extremis*, in the Spaniard's hands. This new clarity may be part of what Eugene brings back from his day's journey through San Francisco.

The plot hinges again, and still on the precipice, the Spaniard suddenly lifts Eugene and twirls him above his head, over the abyss. Eugene does not feel terror, but rather "piloted on great strength." As the Spaniard spins him around, Eugene has another vision: this time his wife Emma is vigorously kissing him and he realizes that by means of the tenderness and mystery of their old sexual passion "there would be a child again."

At this mystical climax of the story, in come the banal and the everyday, for in her prose Welty can swing back and forth between different realms of feeling and meaning with the agility of a poet. She has a marvelous capacity to pivot between the transcendent and the trivial, as though to remind us what mystery can look like from the outside: absurd, strange, silly.

Thus the Spaniard is bellowing and rumbling and holding Eugene up by the knees like a bird. Eugene has his arms out in a sort of ecstasy. This ecstasy, while intensely sexual in its overtones, does not seem to be limited to human intercourse but is rather much richer and deeper in its implications, for the language here is closest to that of religious mystics when they describe union with the divine. And now, while the Spaniard is holding Eugene aloft, a girl calls out from down below where she is strolling with her sweetheart: "Aren't you ashamed of yourself, teasing a little fellow like that, scaring him?"

This breaks the spell. The Spaniard puts Eugene down, they help each other scramble up the cliff, and they finally go to a café where Eugene spends his last penny (except for a streetcar token) on their coffees. The men part almost formally, still with no common spoken language, and the Spaniard is left standing "alone, on a dark corner at the edge of the city . . . looking in the sky for the little moon."

Eugene has climbed up through the trapdoor of his involvement with the Spaniard. And only now, after the whole day's adventures in the Land of the Dead, can he scramble back up through that initial trapdoor of having slapped his wife on such an anniversary. He returns home, racing up the stairs to his flat, and finds Emma with her best friend, Mrs. Herring. Over chowder for dinner, Eugene mentions off-handedly that he has seen the Spanish musician. Emma tells Eugene the man's name—Bartolomeo Montalbano—and the two middle-aged women proceed to gossip, turning the exotic stranger into a man of commonplace maladies and misbehaviors, one who has indigestion, who needs a haircut, who laughs inappropriately in church. Absent, for these women, are the Spaniard's mystery and transcendence, his citizenship in, or dominion over, the World Below. Only Eugene, who is ready for it, only "the man who knows," as Parmenides calls the initiate, can travel to out of his normal world to encounter his spirit guide and explore Nighttown, in San Francisco, in broad light of day.

A Trapdoor in Nighttown

Of course, real life is full of trapdoors—actions that once they are performed seem infinitely hard to undo. One of the strangest such situations I ever observed was in New Orleans, in the mid 1980s.

I was staying in the French Quarter at the Cornstalk Hotel, named for its fence of wrought iron in the shape of corn stalks. Having time to kill in the evenings, I would usually take a walk or go and listen to music, but one night I decided to eat in a restaurant in the Quarter that had an upstairs balcony, overlooking a crossroads, where I could sit and watch the people below without being seen.

As I started dinner, a tall woman sauntered purposively up the street. She wore high-heeled cowboy boots and jeans and a white starched shirt and a cowboy hat and a long blue scarf around her neck. At the corner beneath my balcony, she began to dance, either to some interior music of her own, or a strain of those musics that escape into the night air from every door and window, from the very seams of the buildings, of the Quarter. She was a gorgeous creature, lanky and compelling. As cars prowled down the street, she would beckon to them to slow down, then she slithered up onto their hoods, gyrating and doing improbable things with her long legs, until finally she would slide off the other side of the car. Then she would dance in and around and on top of the bollards that blocked cars from entering one of the cross streets. A crowd gathered to watch such rare choreography, with its masterful invention and control. She would flirt with everyone and then eye the guys and stroll over to them. Then they would look blessed and chosen, especially if she put her silky blue scarf around them and pulled them into the crossroad to dance with her. And the crowd grew and loved her. The crowd was pulsating and would have given her anything.

Perhaps somebody said something, perhaps somebody begged her, or perhaps she just felt like it . . . she started a new sort of dance, throwing her blue scarf over a lamppost. She flung her black cowboy hat onto a bollard. Still she danced, undoing her starched white shirt, one breathtaking button at a time. The crowd grew silent. The people belonged to her. She could have asked anything of them.

I never saw it coming, the hinge on which everything pivoted, though you will have by now. This was 1985 or so, and I was innocent, for I had been studying insects, silk moths, and they did not do things like this. Or rather they did, but when they cast off their skins, there was another silkworm inside, unless they were undergoing metamorphosis. Perhaps that is what the New Orleans street dancer was doing, for when she finally threw off her white shirt, everything changed. Now she tried to dance with the men who had loved her only moments before. But that was when she had been a *she*, and now it was clear that she was not. A collective shudder went through those who had been so captivated only moments before; now no one would accept her gyrations, even when she tendered her blue scarf, taking it from the lamp post and holding it out to them, supplicating.

Soon she put her shirt on again, and soon retrieved her hat, but when she turned it over and held it out for donations no one would even catch her eye. As one, the crowd recoiled, then parted, dispersed, seeped away.

When no one gave her anything at all she put on her hat, whipped her scarf around her neck, and walked with that brilliant androgynous feline step off into the darkness of the Quarter.

What I had witnessed was, of course, a trapdoor, not a hinge, for once the revelation had taken place, there was no way of undoing the collective knowledge that she was not what she had seemed.

The next morning I awoke to gunshots just outside my window. Two cabbies were arguing over a fare; one had pulled a small silvery handgun and was still shooting into the hood of the other's taxicab. Another trapdoor: autocide. As I watched the punctuated cab bleeding its fluids all over the pavement, the hotel maid came into my bedroom with biscuits and coffee. She told me not to worry about a thing.

V | Hinges of the Mind and of the Heart

Hinges and Creative Work and Demons

We have all sorts of hinges in our minds, particularly as we try to settle into creative work. In our present world of finding ethereal connectives everywhere we turn, with internet webs sticky as gossamer, although we might crave a trapdoor into the oubliette of deep attention, getting into work often takes the form of pivoting in and out of focus, as we flutter on the threshold, pursued and tempted and taunted by demons.

For me, some of the most alluring demons are those of light and those of cooking. These are Italianate demons I am talking of here—attractive mischiefs, rather than the Germanic demons, who are carriers of nightmares. I will not talk of those just now.

Some of the sweet demons of light arrive just before evening, when a green sky comes over the garden and the cornfield beyond, with high gold clouds slow and luminous, and those low smoky barracuda clouds scudding towards the east. Green sky is clearly the most wondrous form that sky can take, and I have to watch it even though all afternoon when the light was golden and dancing I found myself swinging from vantage point to vantage point to pay proper attention to it, to decide first of all, is it better here? Or here? And how about now, facing this direction?

Is this what the hummingbird does? We think that he is after nectar, but could he, too, be after light?

❧ ❧

If I had really been working on my manuscript, or if I had been deep into reading the poems of my friend, the chowder would not have been half as good as it turned out to be. But there I was, alone, sitting on the porch, a story I was writing on my lap, thinking, "What if I chop some onions and garlic while I think about the theological differences between my religious mystic and my Jesuit priest." And then, while chopping, "Oh, that's such an easy problem, it's really such a clear distinction, it's simply that the words I have on the page are not the right ones—but smell this purple garlic still shaggy with dirt from the garden." So the garlic demons lured me, and I shuttled in and out of concentration. There were further demons in the just-picked tomato, the chicken jalapeño sausage, the zucchini, the lemon thyme and basil and oregano, the corn grilled in its husk, the leftover lobster.

Earlier that same day, before lunch, I had needed to know how to get the horn off a ram. Can you just snap it off? He is dead, the ram in my story, ritually butchered for the Islamic feast of the Sacrifice. But will a knife do for his horns! Would you need a machete? When I should have been concentrating on the ram's horn, I was beset by the picnic demons who told me that the sandwiches for the beach had to be made from last night's salmon and grilled lemons and a couple of thermoses of the darkest coffee, one hot for him, one iced for me. But that magnificent coiled ram's horn? How do you get it off the animal? (I learned, some months later, that one needs a saw, preferably electric, to do this.)[1]

Only in writing it all down do I see how abducted I get: I spend so much time flickering on the edge of work but not quite in it, the demons of the world of matter and of light, the demons of the garden, and the demons of dinner are always struggling for possession.

I complain about the oscillations, but in fact I am also delighted by them. Edges delight me. Borders and thresholds—these are the terrifying places where I am most at home even while I find them puzzling, doubt-engendering, and loaded with possibilities of choice and danger of permanent exile. My favorite people inhabit the margins. Dawn and twilight make me shudder.

☙ ❧

Hinges are there, too, in the toggling, the flicking back and forth, of changing your mind. In the pivoting of indecision, making up your mind can feel like a welcome plummeting through the trapdoor.

Life is full of the trapdoors of experience, which often stem from our own doings or omissions: "If only I hadn't. If only I had." These are the unchangeables of eternal regret. The things to avoid while we live, so they will not plague us on our deathbeds.

The hinge does not lead to these eternal regrets, because with the hinge one can easily swing back the other way, if one does not like what is going on in the room one has just entered. Hinging is oscillation; trapdooring is more permanent, a jumping down or falling in. But I do not think that all trapdoors are negative. There are wildly positive forms of mental trapdoors. Learning. Knowledge. Understanding. Love.

Once you have really learned a foreign language, you cannot *not* overhear the conversation next to you. Although the stages of learning can involve shimmers and oscillations more like hinges, once you really know something, you know it and, short of cognitive lesions, there is no going back.

This is why the Messenger is generally not a hinge but a form of trapdoor. Consider a mystic vision like the Annunciation, for example. The angel Gabriel comes and tells you something and you are never the same. The telegram used to have the same function. So, often, does the telephone, particularly when it rings in the middle of the night.

Falling in love is a trapdoor. That is why we say *falling*. So, too, may be falling out of love.

And Saw That It Was Good

One morning in my kitchen garden, having crouched and scrabbled in the dirt for hours, I finally stood up and got the hose and watered it all. And as I stood there watching the dust take on the color of earth, I saw that it was good. It was not only fine but it partook of the Good.

Suddenly I felt less guilty for being in the garden with my tomato demons and my lettuce demons and my yard-long-bean demons, rather than at my desk—because here was such an abundance of goodness.

It occurred to me that the phrase that I was echoing might imply something about the god of Genesis. That when God stepped back and "saw that it was good," he was doing what the creative artist does. You would think, or I would, that a *supreme* being would have no question about his workmanship, his art. I mean, this is a Being who may not even have attributes. Well, not physical ones, anyway. And yet each time, after each day's creative work, he sees "that it is good." Actually, *almost* each time: on the second day he neglects to see that it is good; Jewish legend says that the reason for this lack is that on this day he not only makes divisions and separations of the waters above and below by the firmament, but he also creates fire, and the angels, and Hell itself.[2]

On the sixth day, however, things are *very* good. And we have seen earlier what he does during that last dusk of creation.

But here is what I am really saying: in order for God to notice "that it was good," in order for him to see that, he had to be *looking*; that is, there had to be some question for him about the goodness of creation, its success. The text of Genesis does not say, "And God took pleasure that it was good"—though of course he must have done—but rather it feels as though he does the active thing of seeking out whether or not it is good. That "or not"—that self-doubt and necessity of judgment—is what poets and writers, all artists, do. As well as the taking pleasure.

Another way of saying it: God should not have to see that it is good. That should go without saying. The fact that it has to be said indicates something, indicates the presence of doubt. What is a supreme being doing having this kind of doubt? Perhaps it is to give us a model for how to do art. For we, too, hinge between creating, and stepping back to see what we have done. Only once we are outside the work, and not lost in among its parts, can we question and know its goodness.

The Hinge of the Act

When looking at hinges and axles and trapdoors, we are witnessing not just the groan of the liminal but also the necessity of motion itself—in architecture surely, in the cosmos, and also in thought. It suddenly feels as though we have discovered the Verb. Often when we talk of motion, we are thinking of linear motion going from here to there, towards a goal. But perhaps the more primal, more mysterious sense of motion is that of oscillating, rotating, turning around a still core—seemingly unchanging yet ever versing and reversing, somehow moving forward, hinging into rooms beyond.

Figure 25. *Girl Sleeping* by Johannes Vermeer (see Plate X).

In Vermeer's painting *Girl Sleeping*, or *Fille Assoupie*, the girl, or young woman, sits at a table that is covered with both a fringed tablecloth and an oriental carpet (Figure 25). Her eyes are closed, but despite the title, I do not think that she is sleeping. She ignores the platter of fruit and the carafe of liquid in front of her as she relaxes, forming a diagonal, her slightly bowed head propped up by her right hand. Her left hand rests on the carpeted table. The chair beside her is empty, the absence of another freeing her to explore within. On the wall hangs a painting but it is hard to see what it depicts, though it is thought to contain Cupid, whose leg can just be seen, along with a mask. Behind the girl, an open door leads to the room beyond. This is more illuminated, with a drop-leaf table, another painting or a mirror, and what looks like a window because of its heavy wooden frame.

It is as though as we gaze on her, in her reverie, we are allowed only hints of what she is imagining, in that room beyond consciousness. She is too pensive to be sleeping. The mood of utter quiet—despite the intricate texture and color of the carpet she touches—shows that she is daydreaming, imagining. For her, that chair in front of her is no longer empty, but filled with the presence of the someone we cannot yet see, for she is still writing him into existence. That this occurs in an ordinary room surrounded by ordinary things, chairs and tables and food and drink, indicates that incubation can happen anywhere that one can achieve the necessary stillness.

❦ ❧

The great paradox of writing is that one puts oneself apart from people for large stretches of time in order, finally, to talk to them. In order to talk to others, one has to avoid them and listen to one's own voices for a while. Some of these voices can be subtle and elusive, others completely skittish and perhaps otherworldly. One needs silence and stillness to hear them. The pivoting back and forth between companionship and solitude is one of the weirdest things we do. Fleeing external lives and loves, we finally inhabit the images and music and meaning of what we are writing until somebody calls: "Where is the cat? Have you seen the cat?" And so we get up and let the cat in, for it is raining now, and we close some windows and go back to the room of the imagination where the damp cat has crept into our writing and the rain drums on the skylight as we turn the blue glass paperweight over and over in our hands. How we maneuver between these worlds, how we take the strands of life and longing and turn and weave them into what we write, and how our writing, made from contemplation of that desire and that absence, in turn constructs the self and informs the soul—these are the crucial hinges. The axial aloneness of the writer at the still center, calling the body, and the book, and the reckoning of the heavens into being.

Notes

Chapter I

1. From *Ancient Engineers: Technology and Invention from the Earliest Times to the Renaissance* by L. Sprague de Camp; page 54.

Chapter II

1. I am deeply indebted to Philip Fisher, who first told me, in a chance meeting in Harvard Yard several years ago, of these notions of liminality and entrancement in the beginnings of early novels, including *Remembrance of Things Past* and *Alice in Wonderland*.

2. From the first page of *Alice's Adventures in Wonderland* by Lewis Carroll.

3. The source for this discussion of Hermes, and for all my discussions of *The Odyssey* is the brilliant *Homeric Moments* by Eva Brann.

4. For a compelling and well-documented discussion of the practice of incubation and this aspect of Apollo, see Peter Kingsley's *In the Dark Places of Wisdom*, pages 79–92. See also note 13 below.

5. Ibid., Kingsley, pages 102–104.

6. See Diogenes Laertius, *Lives of Eminent Philosophers*, Vol. II, page 339, 8.21. Diogenes is quoting Hieronymus of Rhodes, who lived around 300 BC and was a disciple of Aristotle.

7. See Benjamin Foster's *The Epic of Gilgamesh*, pages xi–xii.

8. Andrew George has done the best translation of Gilgamesh that I have found. My discussion is totally indebted to his introduction and notes.

9. For a stunning view of contemporary Mesopotamian boats, poled and otherwise, see the movie *Zaman: The Man from the Reeds*, filmed in Iraq in 2003.

10. *Odyssey* X: 561; in Fitzgerald's translation, pages 180–181.

11. Eva Brann, *Homeric Moments*, page 205.

12. Bruno Schulz, *Sanatorium under the Sign of the Hourglass*, page 47.

13. Ibid., page 48.

14. Ibid., page 48.

15. Ibid., page 48.

16. Kingsley's translation of the Proem is taken from *In the Dark Places of Wisdom*, which is also my source for his arguments about Parmenides as healer-prophet. In my opinion, these arguments, while disputed, are much too well researched, and also too conducive to thought, to be ignored. For the other two sections of the long poem of Parmenides, the best translation I have found is in David Gallop's *Parmenides of Elea: Fragments*.

17. Although she is unnamed in the text, Kingsley notes that because she is Queen of the Dead, Persephone is very rarely called by name in Greek texts, and that in this context, when she is referred to simply as "the Goddess," it is understood to be Persephone.

18. According to Kingsley, the use of the word *chasm* is indicative that we have crossed into Tartarus or Hades, and have thus made a descent.

19. From "Meditation at Lagunitas" in *Praise* by Robert Hass. Emphasis is mine.

20. "The Voice You Hear When You Read Silently," from *New and Selected Poems* by Thomas Lux.

21. For a brilliant discussion of authorial instruction and solidity in fiction, see Elaine Scarry's *Dreaming by the Book*, particularly the chapters "On Vivacity" and "On Solidity."

Chapter III

1. Ovid, *Metamorphoses, Book X;* in Charles Martin's translation, page 3.

2. Virgil, *Georgics* 4; in David Ferry's translation, page 9.

3. Of course, that incline was a very long one, according to Parmenides, "as far as longing can reach."

4. *The Georgics of Virgil*, translated by David Ferry, pages 177–181.

5. Note that the verbs *to touch* and *to taste* do not have different transitive and intransitive forms in English. Perhaps this is because touching and tasting do not happen at a distance, but the object of examination is in contact with the tongue or the skin. With the sense of smell, things can happen at a distance, although *to smell* has no intransitive form in English either.

6.   As when someone planning a journey prepared a lamp,
the gleam of blazing fire through the wintry night,
and fastened linen screens against all kinds of breezes,
which scatter the wind of the blowing breezes
but the light leapt outwards, as much of it was finer,
and shone with its tireless beams across the threshold;
in this way [Aphrodite] gave birth to the rounded pupil,
primeval fire crowded in the membranes and in the fine linens.
And they covered over the depths of the circumfluent water
and sent forth fire, as much of it was finer.

From *The Poem of Empedocles* by Brad Inwood, page 249: fr. 103 by Inwood's numbering. The "[Aphrodite]" is not in the Greek, but he must be supplying this from fr. 86: "From which divine Aphrodite fashioned tireless eyes"; fr. 87: "Aphrodite wrought [them] with the dowels of love"; and fr 95: "When they first grew together in the devices of Kupris [i.e., Aphrodite]."

7. Aristotle, *Sense and Sensibilia*, 437b1-23; cited in Peter Pesic's article "Seeing the Forms" in *Journal of the International Plato Society*.

8. This and the discussion of Plato that follows are taken from Pesic's "Seeing the Forms."

9. Ibid., Pesic, page 6. As Pesic notes, Plato's "understanding of vision itself critiques the kind of thoughtless viewing in which inner and outer fires do not adequately engage."

10. Ibid., Pesic, page 3.

11. Ian Fisher, "Italy: Men Can't Grope . . . Themselves," *The New York Times*, February 28, 2008:

> Whatever their reason might be, a passing hearse or simple discomfort, Italy's highest court ruled that men may not touch their genitals in public. The ruling settled an appeal by a 42-year-old worker from Como, north of Milan, who was convicted in May 2006 of "ostentatiously touching his genitals through his clothing," though his lawyer argued it was a problem with his overalls. But the court struck against a broader practice: a tradition among some Italian men of warding off bad luck by grabbing the crotch. The court ruled that this "has to be regarded as an act contrary to public decency, a concept including that nexus of socio-ethical behavioral rules requiring everyone to abstain from conduct potentially offensive to collectively held feelings of decorum." The judges suggested that if they need to, men can wait and do it at home.

12. Alan Dundes, "Wet and Dry, the Evil Eye: An Essay in Indo-European and Semitic Worldview," in *The Evil Eye: A Casebook* (University of Wisconsin Press, Madison, 1981) pages 257–312.

13. Thus the combination of the notion of envy with the importance of fluids and the idea of limited good can explain the widespread custom of toasting others when we drink: Here's to your health. "I drink, but not at your expense. I am replenishing my liquid supply, but I wish no diminution in yours." (Dundes, page 268.). The custom of wishing other diners "*Bon appetit*" in France or "*Buen provecho*" in Spain is similarly a likely way of reassuring the other person that he should not fear your food envy. Since a waiter in a restaurant may also be prone to envying the one who dines, he is given a tip, which most likely comes from the word *tipple* since in many European languages the word for *tip* contains the word for drink, as in *pourboire* and *Trinkgeld*, again inviting the waiter to drink, to thus equalize the distribution of life fluids.

14. Emily Dickinson #1129. Available at http://en.wikisource.org/wiki/Tell all_the_Truth_but_tell_it_slant_%E2%80%94.

15. I am grateful to Zeke Mazur for pointing this out to me.

16. Laura Hendrie, personal communication.

17. See Apollodorus, *The Library*, 2.4.2.

18. Ovid, *Metamorphosis*, Book IV, 1010–1026.

19. *Brewer's Dictionary of Phrase and Fable*, page 1124.

20. Frazer, *The New Golden Bough*, page 92.

21. Rosemarie Bernard, personal communication. For more on this and related topics, see "Ise and the Modern Emperors" by Rosemarie Bernard, in *The Emperors of Modern Japan*, edited by Shillony, Ben-Ami (Leiden, EJ Brill, 2008).

22. Official website of the Ise Shrine is http://www.isejingu.or.jp/english/myth/myth4.htm.

23. Jamaica Kincaid, personal communication.

24. Ovid, *Metamorphosis*, Book III, 330–403.

25. Zeke Mazur, "*Unio Magica:* Part 1: On the Magical Origins of Plotinus' Mysticism," *Dionysius*, Vol. XXI, December 2003, pages 23–52.

26. For the full tale of Cupid and Psyche, see *The Golden Ass* by Apuleius.

27. Alter, *The Five Books of Moses*, page 503–506.

28. Exodus 33:21–24.

29. Exodus 33:23.

30. Marie-Rose Séguy, *Mirâj Nâmeh: The Miraculous Journey of Mahomet*, Bibliothèque Nationale Paris Manuscrit Supplément Turc 190, New York, Braziller, 1977.

31. Schneemelcher's *New Testament Apocrypha: Volume One—Gospels and Related Writings*, pages 537–553.

32. Herbert Dandy, *The Mishnah*, Oxford, Oxford University Press, 1933, pages 212–213; cited in Barry Mazur's *Imagining Numbers (Particularly the Square Root of Minus Fifteen)*, page 241.

33. Ovid, *Metamorphoses*, Book X, 83–85.

34. Joyce Carol Oates has a two-page story, "The Blue-Bearded Lover," about the wife who does *not* look in the secret room. My reading of the story is that the ridiculous vapidity of the wife highlights the risible premise of the story.

35. Pesic, "Seeing the Forms," page 10.

36. Gregory Nagy, "Can Myth Be Saved?" pages 240–248.

Chapter IV

1. The description and analysis of this icon is taken from the article "Russian Iconography to the Beginning of the XVI Century" (in Russian), by V. N. Lazarev, http://www.icon-art.info/book_contents.php?book_id=7&chap=8&ch_l2=27#pic128, September 12, 2008. I am grateful to Sasha Makarova for her translation.

2. St. Gregory of Nazianzus (329–430) believed that the unbaptized would neither be punished nor access the full glory of God. But St. Augustine of Hippo (354–430) said that baptism was necessary for salvation and that even babies would go to Hell if not baptized. St. Thomas Aquinas (1226–1274) was the first major theologian to speculate about the existence of a place called limbo. See "Catholic Church buries limbo after centuries," *ReligionNewsBlog*, http://www.religionnewsblog.com/18025/limbo, April 21, 2007.

3. Christ's Descent to Hell is mentioned in The Apostolic and the Athanasian Creeds—it is not mentioned in the Nicene, nor the Chalcedonian.

4. For a discussion of the possible date of this Gospel, as well as the text of various Greek and Latin versions, see "Christ's Descent into Hell" in The Gospel of Nicodemus in New Testament Apocrypha: Volume One, edited by Wilhelm

Schneemelcher, pages 501–536. For the text of Satan's conversation with Hades and Christ's entrance into Hell, see pages 523–526.

5. Genesis 32:24–26: "Then Jacob was left alone; and a Man wrestled with him until the breaking of day. Now when He saw that He did not prevail against him, He touched the socket of his hip; and the socket of Jacob's hip was out of joint as He wrestled with him. And He said, "Let Me go, for the day breaks." (Authorized (King James) Version.)

6. Robert A. Kraft, "The Codex and Canon Consciousness," pages 229–233.

7. David K. Lynch and William Charles Livingston, *Color and Light in Nature*, pages 38–39.

8. Along a different line of thought, for a marvelous exposition of the power of twilight and what it does to vision of all kinds, see Pete Turchi's lecture "Unresolved Characters," given at the residency of the MFA Program for Writers at Warren Wilson College in July 2005.

9. *The Soul of the Ape* by Eugene Marais, pages 101–106.

10. *The Book of Legends (Sefer Ha-Aggadah)* by Hayim Nahman Bialik and Yehosua Hana Ravnitzky, page 16.

11. Ibid., page 125.

12. See *The Art of the Poetic Line* by James Longenbach.

13. Rayor, *Homeric Hymns*, page 17. I am grateful to Greg Nagy for pointing out to me this early instance of the trapdoor in literature.

14. Ibid., page 31.

Chapter V

1. Stelios Moschapidakis, personal communication.

2. See Louis Ginzberg, *The Legends of the Jews: Volume 1*, pages 15–16. Ginzburg gives a wonderful description of the hells created on the second day:

> Hell has seven divisions, one beneath the other. They are called Sheol, Abaddon, Beer Shahat, Tit ha-Yawen, Sha'are Mawet, Sa'are Zalmawet, and Gehenna…. Each of the seven divisions in turn has seven subdivisions, and in each compartment there are seven rivers of fire and seven of hail. The width of each is one thousand ells, its depth one thousand, and its length three hundred, and they flow one from the other and are supervised by ninety thousand Angels

of Destruction. There are, besides, in every compartment seven thousand caves, in every cave there are seven thousand crevices, and in every crevice seven thousand scorpions. Every scorpion has three hundred rings, and in every ring seven thousand pouches of venom, from which flow seven rivers of deadly poison. If a man handles it, he immediately bursts, every limb is torn from his body, his bowels are cleft asunder, and he falls upon his face. There are also five kinds of fire in hell. One devours and absorbs, another devours and does not absorb, while the third absorbs and does not devour, and there is still another fire, which neither devours nor absorbs, and furthermore a fire which devours fire. There are coals as big as mountains and coals as big as hills, and coals as large as the Dead Sea, and coals like huge stones, and there are rivers of pitch and sulphur flowing and seething like live coals.

Time Line

List of Figures

from a Lucanian red-figured calyx-krater. Bibliothèque Nationale de France, Paris, France. Photograph by Marie-Lan Nguyen. Photograph courtesy of Wikimedia Commons.

11. *Orpheus and Eurydice with Hades and Persephone* by Peter Paul Rubens (1577–1640). Prado Museum, Madrid, Spain. P1667. Photograph copyright Museo del Prado – Madrid – Spain.

12. Persephone and Hades (c. 440–430 BC), attributed to the Codrus painter. Tondo of an Attic red-figured kylix. Said to be from Vulci. British Museum, London, UK. Photograph by Marie-Lan Nguyen. Photograph courtesy of Wikimedia Commons.

13. Perseus, Athena holding Gorgon head, and Hermes by the Taporley painter. Mixing bowl (bell krater). Greek, South Italian, Classical Period, about 400–385 BC. Place of Manufacture: Italy, Apulia. Ceramic, Red Figure. Height 30,5 cm (12 in.). Museum of Fine Arts, Boston. Gift of Robert E. Hecht, Jr. 1970.237. Photograph © 2010 Museum of Fine Arts, Boston.

14. *Creation of the Sun, Moon, and Planets* by Michelangelo Buonarotti (1475–1564). Fresco. Sistine Chapel, Vatican, Rome. Photograph courtesy of The Yorck Project/Wikimedia Commons.

15. "The American Cowslip" (*Dodecatheon meadia*) by Dr. Robert John Thornton (1768–1837). From *Temple of Flora* (1801). Engraving. Photograph by Paul Horowitz.

16. *Anastasis: Christ Saving Adam and Eve.* Byzantine fresco, early 14th century AD. Chora Church (Kariye Camii), Istanbul. Photograph by Gunnar Bach Pedersen. Photograph courtesy of Wikimedia Commons.

17. *Descent into Limbo* by Dyonisius (1440–1510). Russian icon from the Ferapontov Monastery. Tempera on wood. Russian State Museum, St. Petersburg, Russia. Photograph courtesy of Wikimedia Commons.

18. *The Descent into Limbo* by Duccio (di Buoninsegna) (c. 1260–1319). Panel from the back of the Maestà altarpiece. Tempera on wood. Museo dell'Opera Metropolitana, Siena, Italy. 1308-11. Photograph © Scala/Art Resource, NY.

19. *Christ's Descent into Limbo* by Jacopo Bellini (c. 1400–1470). Musei Civici, Padua, Italy. Photograph © Cameraphoto Arte, Venice/Art Resource, NY.

Bibliography

Alter, Robert. *The Five Books of Moses: A Translation with Commentary.* New York: Norton, 2004.

Alwan, Amer (director). *Zaman: The Man from the Reeds.* Issy-les-Moulineaux: Arte France, 2003.

Apollodorus. *The Library,* two volumes, translated by Sir James George Frazer. Cambridge, MA: Harvard University Press, 1921.

Bataille, Georges. *Lascaux or the Birth of Art.* New York: Skira, 1955.

Baxter, Charles. *The Feast of Love.* New York: Pantheon, 2000.

Bialik, Hayim Nahman and Yehosua Hana Ravnitzky. *The Book of Legends (Sefer Ha-Aggadah),* Annotated Edition, translated by William G. Braude. New York: Schocken, 1992.

Brann, Eva. *Homeric Moments: Clues to Delight in Reading the Odyssey and the Iliad.* Philadelphia: Paul Dry Books, 2002.

Brewer, E. Cobham. *Brewer's Dictionary of Phrase and Fable,* 17th Edition. New York: Collins, 2005.

Carroll, Lewis. *Lewis Carroll: Complete Works.* New York: Vintage, 1976.

de Camp, L. Sprague. *Ancient Engineers: Technology and Invention from the Earliest Times to the Renaissance.* New York: Barnes and Noble Publishing, 1990.

Dickinson, Emily. "Tell all the Truth but tell it slant –." *Wikisource.* Available at http://en.wikisource.org/w/index.php?title=Tell_all_the_Truth_but_tell_it_slant_%E2%80%94&oldid=354377 (posted March 11, 2007) .

Diogenes Laertius. *Lives of Eminent Philosophers,* Vol. II, translated by R. D. Hicks. Cambridge, MA: Harvard University Press, 1942.

Dundes, Alan. "Wet and Dry, the Evil Eye: An Essay in Indo-European and Semitic Worldview." In *The Evil Eye: A Casebook*, edited by Alan Dundes, pp. 257–312. Madison: University of Wisconsin Press, 1981.

Fagles, Robert (translator). *The Odyssey*. New York: Penguin, 1999.

Fisher, Ian. "Italy: Men Can't Grope . . . Themselves." *The New York Times*, February 28, 2008.

Fitzgerald, Robert (translator). *Homer: The Odyssey*. New York: Farrar, Straus and Giroux, 1998.

Foster, Benjamin. *The Epic of Gilgamesh*, Norton Critical Edition. New York: Norton, 2001.

Fox, Paula. *The Widow's Children*. New York: Norton, 1999.

Frazer, Sir James George. *The New Golden Bough*, edited by Theodor H. Gaster. New York: Doubleday/Anchor Books, 1961.

Gallop, David. *Parmenides of Elea: Fragments*. Toronto: University of Toronto Press, 1984.

George, Andrew. *The Epic of Gilgamesh: A New Translation*. London: Penguin, 1999.

Ginzberg, Louis. *The Legends of the Jews: Volume 1—From the Creation to Jacob*. New York: Cosimo, 2005. (Originally published in 1909.)

Hass, Robert. "Meditation at Lagunitas." In *Praise*. New York: Ecco, 1990.

The Holy Bible. Authorized (King James) Version. Nashville: The Gideons International, 1964.

Inwood, Brad. *The Poem of Empedocles*. Toronto: University of Toronto Press, 1992.

Kingsley, Peter. *In the Dark Places of Wisdom*. Inverness: The Golden Sufi Center, 1999.

Kraft, Robert A. "The Codex and Canon Consciousness." In *The Canon Debate*, edited by Lee Martin McDonald and James A. Sanders, pp. 229-233. Peabody MA: Hendrickson, 2002.

Lattimore, Richmond (translator). *The Odyssey of Homer*. New York: HarperPerennial, 1999. (Originally published in 1967.)

Longenbach, James. *The Art of the Poetic Line*. Saint Paul: Graywolf, 2007.

Lux, Thomas. "The Voice You Hear When You Read Silently." In *New and Selected Poems*. Boston: Houghton Mifflin, 1997.

Lynch, David K. and William Charles Livingston. *Color and Light in Nature*. Cambridge: Cambridge University Press, 2001.

Mansfield, Katherine. "The Garden Party." In *The Garden Party and Other Stories*. London: Penguin, 1983. (Originally published in 1922.)

Marais, Eugène. *The Soul of the Ape*. London: Penguin, 1973.

Mazur, Barry. *Imagining Numbers (Particularly the Square Root of Minus Fifteen)*. New York: Farrar, Straus and Giroux, 2003.

Mazur, Zeke. "*Unio Magica*: Part 1: On the Magical Origins of Plotinus' Mysticism." *Dionysius* XXI (December 2003), 23–52.

Nagy, Gregory. "Can Myth be Saved?" In *Myth: A New Symposium*, edited by G. Schrempp and W. Hansen, pp. 240–248. Bloomington: Indiana University Press, 2002.

Ovid. *Metamorphoses*, translated by Charles Martin. New York: Norton, 2004.

Plato. *Timaeus*. In *Plato: Complete Works,* edited by John M. Cooper. Indianapolis: Hackett, 1997.

Pesic, Peter. "Seeing the Forms." *Journal of the International Plato Society* 7 (2007), Article 7. (Available at http://www.nd.edu/~plato/plato7issue/ Pesic,%20Seeing%20the%20Forms1.pdf.)

Rayor, Diane (translator). *The Homeric Hymns*. Berkeley: University of California Press, 2004.

Ruspoli, Mario. *The Cave of Lascaux: The Final Photographs*. New York: Harry N. Abrams, 1987.

Scarry, Elaine. *Dreaming by the Book*. New York: Farrar, Straus and Giroux, 1999.

Schneemelcher, Wilhelm (editor). *New Testament Apocrypha: Volume One—Gospels and Related Writings*, Revised Edition, translated by R. McL. Wilson. Louisville: Westminster John Knox Press, 2003.

Schulz, Bruno. *Sanatorium under the Sign of the Hourglass*. Boston: Houghton Mifflin, 1977. (Originally published in Poland in 1937.)

Séguy, Marie-Rose. *Mirâj Nâmeh: The Miraculous Journey of Mahomet*, Bibliothèque Nationale Paris Manuscrit Supplément Turc 190. New York: Braziller, 1977.

Virgil. *The Aeneid*, translated by Robert Fitzgerald. New York: Random House, 1981.

Virgil. *The Georgics of Virgil*, Bilingual Edition, translated by David Ferry. New York: Farrar, Straus and Giroux, 2005.

Welty, Eudora. "Music from Spain." In *The Golden Apples*. New York: Harcourt Brace Jovanovich, 1947.

Index

Printed and bound by CPI Group (UK) Ltd, Croydon, CR0 4YY

23/10/2024

01777699-0001